SAVE YOUR HORSE!

A Horse Owner's Guide To Large Animal Rescue

Written by

Michelle Staples

And a few friends

Save Your Horse!

A Horse Owner's Guide to Large Animal Rescue

Michelle Staples

Published by: **Red Jeans Ink**
Post Office Box 881 www.redjeansink.com
Buffalo, NY 14201 U.S.A.

The intention of this book is to provide information, helpful material and resources, and does not provide professional services. The author, publisher and all contributors disclaim any and all responsibility for any loss, injury or death which may occur as the result of use of material contained in this book.

Unattributed quotations and photographs are by Michelle Staples. Some of the original artwork is by Sue Duffield.

ISBN #0-9785685-9-1

First Printing, October 2006
Second Printing, July 2007
Third Printing, December 2008
Fourth Printing, June 2010
Fifth Printing, October 2014

Printed in the United States of America

Library of Congress Cataloging-in-Publication Data

Staples, Michelle.
Save Your Horse! A Horse Owner's Guide to Large Animal Rescue/Staples
Includes bibliographical references and index.
ISBN #0-9785685-9-1
1. Animals -- Horses – United States. I. Title.
2. Large Animal Rescue -- Horses
3. Animals – Disaster
4. Animals – Medical
5. Animals – Rescue
6. Disaster – Animals

Cover picture of Aerial being lifted from the mud is courtesy of Dr. Rebecca Gimenez.
Read about Aerial inside.

Acknowledgements

No book is written in a vacuum, especially a technical book.
There are many experts who have reviewed this book to make sure it contains the safest and most accurate information available; others have contributed articles.
Of course, things can go very, very wrong in an incident involving large frightened animals and we do not offer guarantees from harm.
Some of these people offer training and expertise to help keep us as safe as possible under the circumstances.
Their work is heroic.
I'd like to thank the following contributors, listed alphabetically:

Charles Anderson – *Anderson Rescue Equipment*
Dr. Kathleen Becker – *Hast PSC, LAR Instructor*
Ana Bertero – *Contra Costa County Fire Protection District, retired*
Dr. Steve Bohn – *Gualala Veterinary Clinic*
Patrick Aaron Charlesworth – *PC Troubleshooting*
Mark Cole -- *USRider*
Larry Collins – *Eastern Kentucky University*
Timothy Collins – *Collins Rescue Equipment*
Warren Craig – *ASAR Services*
Sue Duffield -- *artist*
Nicole Ehrentraut- *DaVinci Equine Emergency Transport*
Deb and John Fox – *Large Animal Rescue Co. Inc.*
Drs. Rebecca and Tomas Gimenez – *Technical Large Animal Emergency Rescue*
Jim Green -- *Hampshire Fire and Rescue Service, UK*
Cherry Hill – *author of over 30 horse-related books*
Chris Jonason, *Wave Trek Rescue*
Joe Key – *Expert Witness, firearms*
Roger Lauze – *Massachusetts SPCA, Nevins Farm*
Mary Anne Leighton – *Equine Emergency Rescue, Australia*
Laurie Loveman – *Barn Fire Safety Expert*
Kelly Harrington Nilssen – *ASPCA Disaster Response Services*
Vicki Schmidt – *Frandford, ME Fire Dept.*
Allan Schwartz – *HSUS Disaster Services*
Dina Stephens – *Contra Costa County Fire Protection District*
Eric Thompson – *Emergency Equine Response Unit*
Jennifer Woods –*J Woods Livestock Services*
Julie Young – *Niedner, Inc.*

This book is dedicated to my mom.
Still glowing at 98!

I couldn't have a better role model.
I want to be just like you when I grow up, Mom.

Aerial

Although now an almost middle aged lady, and the seasoned veteran of many years of LAR training sessions, Aerial started her path at Thanksgiving in 2003 when she was rescued from neglect by Dr. Rebecca Gimenez. The Paint filly's owner didn't know a lot about horses and told Rebecca that the filly's mother had died when Aerial was about 2 months old, but "she didn't like to eat the grain he offered, so she had gotten a little skinny". He didn't know about replacement milk or foods. The filly had gotten sick so he called the vet out a couple times, but she hadn't gotten any better. He didn't want to spend any more money and was happy to sell Aerial for a dollar.

Aerial went directly to a veterinary clinic where the exam showed leg, spine, and nerve problems, and a Body Condition Score (BCS) of 1.5. She was past the point of caring if she lived or died. She was 12.2 hands and 212 pounds by the weight tape, at eight months old. She went on a "starving horse" diet and filled out to the beauty she is now.

From that first day at the clinic Aerial was taught to lie down on command so that she could be safely triaged, vaccinated, injected and palpated. Occasionally flipping her onto her back allowed her vertebrae to move. Her doctor donated surgeries to correct her club feet.

By Christmas she was allowed to run loose in the stable yard and visit other horses over the fence, and by New Years had a second surgery.

Ground training continued with desensitization, moving, turning, backing up, and learning to yield to pressure. Aerial learned to follow, come when called, load in horse trailers, patiently stand tied, and stand for grooming. Aerial made her first public demonstration at a local ground handling clinic, backing calmly out of the trailer. She fearlessly handled all the other horses, crowds of people, and unusual obstacles. The crowd loved her and a star was born.

Now allowed to live with other horses, she was "ponied" on trail rides, but also followed loose on short rides in controlled conditions to encounter natural obstacles, water, and tight confined spaces.

By May she joined Rebecca's training program. She was perfect at her job, laying down on command, turning over, demonstrating webbing and rope work, and even being lifted off her feet using a crane, all with no sedation.

Aerial has gone on to give many demonstrations around the eastern US. She wears an equine floatation device into water, jumps into a deep mud pit to simulate a rescue scenario (as seen on our front cover), or simulates a broken leg where she has to be removed from the accident scene tied down on a Rescue Glide. She seems to enjoy the attention and will leap into the trailer without a halter, ready to go to the next stop!

TABLE OF CONTENTS

INTRODUCTION

It's been a beautiful day. You placed in two classes at the local horse show, visited with some long-time friends, and now you're headed home. The traffic is heavy and the road crowded with horse trailers.

Suddenly, the traffic stops. You can't see over the slight rise in the road but you can hear the screams of a horse and the sickening sound of hooves pounding on a trailer wall.

You park your truck and turn off the engine. As you climb out you can see other drivers doing the same. You race toward the clamor and as you top the rise, you're stopped by a horse owner's nightmare. There, in the middle of the road, is an overturned trailer.

The scene is chaotic. People milling around; what appears to be the driver sitting on the side of the road, dazed, with blood dripping down her face. The trailer itself is shuddering with the constant pounding of horse hooves, and above it all are the screams of the horse inside.

What can you do? You hear the sirens that indicate the emergency responders are about to arrive. Will THEY know what to do? Probably not. You've heard the horror stories of similar incidents: emergency responders and bystanders injured or killed; horses irreparably injured by enthusiastic but untrained emergency responders. It's scary to think that the very people who are trained to care for you in an accident may not have the training to help your horse.

The book you are holding in your hands can help turn a potentially lethal situation into a success story. Read it, study it, and carry it in your trailer or vehicle.

After taking the Large Animal Rescue (LAR) class taught by Large Animal Rescue Co. Inc. in California, it became very clear that the information taught in the class needed to be made available to ALL horse owners and the official rescue personnel who might be helping them. After three years of research and consulting experts in the field, this book is the result. It contains the steps necessary to successfully extricate a large animal – particularly a horse – from an overturned trailer, out of mud or a ravine, or off a cliff.

SAVE YOUR HORSE! is designed to be used as an on-scene resource for emergency responders who do not understand horses and are unfamiliar with the training available in Large Animal Rescue. Besides helping you get your horse out of a difficult situation, you can hand it over to rescuers on-scene and they can follow the steps to successfully rescue your horse. This book does not, in any way, take the place of the training offered by experts in the field and will not make you an expert. It is offered as an introduction for horse owners to the field of Large Animal Rescue. My advice is – encourage your emergency responders to sign up for classes; sign up for classes and experience the training yourself!

While it's true you will learn methods for rescuing a large animal from reading a book, it's also true that "a little knowledge can be dangerous." There is so much more involved in the actual practice of this kind of specialized work than any book can portray. Even rescuers, who are trained in the basic methods of the extrication of humans, may not realize the scope of the dangers inherent in working with stressed-out, panicked, large animals.

Familiarize yourself with the information and consider carrying some of the rescue equipment, as well as this book, in your vehicle. Emergency responders are much more likely to attempt to rescue your horse if they can be shown there are methods that ensure their own safety. This work is technically challenging and very dangerous. It should not be attempted by untrained owners but should be left to the emergency responders whose job it is to do this kind of work.

While the safest possible methods for various types of rescues are presented herein, there is no guarantee that all will go well and no one will get hurt and there is no "right way" that works for every situation. When using this book, **always** be prepared for unexpected mishaps and, as all responders are trained to do, work with a buddy.

SECTION ONE

SECTION ONE details step-by-step procedures for extricating your horse from the most common life-threatening situations. It gives specific information on how to use ropes and straps to get the job done safely. At the beginning of the section is an introductory letter addressed to the person directing the rescue effort at the scene of the incident (Incident Commander). It explains why he or she should use these procedures, details who should be called, and who is responsible.

Also included is information on rescuer safety to help allay some of the instinctual fears associated with working with large animals. As horse owners we take for granted working in the same space as our large animals, but we need to be aware of the danger that familiarity breeds and not become complacent to the point of endangering our lives. This will not be the case with those brave women and men who come to your rescue. Ole Dobbin may be your best friend at home, but one thousand pounds of flailing, panicked horseflesh will definitely be cause for concern to someone unfamiliar with your friend's normal pussycat demeanor.

Some emergency responders may be reluctant to get involved with your animals, saying their job is to protect people not critters. Their job is also to protect property. If rescuing animals is not "part of the job description", the justification of saving your property may be within their purview. Horses are a multi-billion dollar industry all over the world. Suddenly, equines take on new value!

To aid responders in working with your non-equine large animals, Jennifer Woods of J Woods Livestock Services has written a chapter on responding to livestock incidents. Jennifer has many years of experience involving hundreds of responses and is the foremost expert in the field.

The best way to help your emergency responders do THEIR job is to do your best to prevent an incident from developing.

It's amazing how many new horse owners will take to the highways hauling a horse trailer with no idea of the skill involved in doing so.

Ask any truck driver; it's not a matter of getting behind the wheel of your vehicle and pulling out into traffic. You need to understand the mechanics of your vehicle and how it will react in an emergency. You need to know about the animals you are hauling – why they do what they do, and when to expect a negative reaction. And you need to practice without your animals in the trailer before you ever let them set hoof inside. If possible, get into the box of the trailer and have a friend drive you around your property. You'll be amazed by the roughness of the ride. When you can't see where you're going you can't brace for turns or stops. A five minute ride will give you new understanding of the stress traveling will put on your horse.

Another way to make your travels safer is to prepare your horse. Don't be fooled by a horse's age. Just because your horse is mature doesn't mean he's an experienced traveler. Teach him how to ride in a trailer. The few hours you spend up front with your horse will reap you great benefits as you travel.

"But," you say, "I never trailer my horse. I don't even OWN a trailer!" Here are a couple of "what ifs" that might convince you otherwise. What would happen if you were forced to evacuate from your home or boarding stable? There would be no time to politely ask your horse to get into a stranger's trailer, and wait for him to acquiesce. If you are not the first one to be picked up, would your horse load into a strange trailer with a horse he doesn't know? How well would he travel in the "stop and go" traffic of an evacuation?

What if your horse were really sick or injured and you needed to transport him to a veterinary hospital? Some situations need more experience or equipment than a field diagnosis or field surgery and are beyond the means of your local veterinarian.

Since not all incidents involve a trailer, how else can you be prepared?

Look around your property with an eye to danger. Do you have an abandoned well or septic tank on your property? Horses, other animals, and children have all been victims of these hazards. If you can't fill in a well to make it safe, pour a concrete cap to cover it. Wood rots and, while appearing to be intact, can crumble under even the smallest weight.

Is there a swimming pool that would be an "attractive nuisance" to children and animals alike? There have been many cases of fleeing deer jumping a fence and ending up in pools. Likewise, horses, cows, moose, and smaller critters such as children, sheep, dogs and pigs have fallen victim to the slippery sides and depth of water of in-ground pools.

Coverings are not adequate protection. Any significant weight will pull the cover under water, and then the unlucky victim has to contend with that as well as the water! Make sure your pool is adequately protected by fencing even if you live in the country. If your pool is not covered, float a board in the water for unfortunate smaller critters to climb onto so they don't drown. We floated a board in our horse trough after discovering a very dead squirrel spread-eagled in it!

Another hazard in many areas of the world is quicksand. If you live in such an area, learn how to extricate yourself and your animals. Remember the old westerns from Saturday morning TV? Someone – usually the bad guy – would fall into quicksand and slowly sink out of sight. On the "good guy" side, there was always time for Lassie to get a branch or alert a kindly stranger if his master, Timmy, became mired in the stuff. Like Timmy, you have time to get out, contrary to movie thrillers. The most important and seemingly obvious thing to do is breathe and don't panic.

According to the *How Stuff Works* (www.howstuffworks.com) website, quicksand is usually no deeper than a few feet and consists of sand that is supersaturated – "a mushy mixture of sand and water that can no longer support any weight." It doesn't pull you down, but if you thrash in it your own weight will cause you to sink. "The worst thing to do is to thrash around in the sand and move your arms and legs through the mixture. You will only succeed in forcing yourself farther down into the liquid sandpit. The best thing to do is to make **slow movements** and bring yourself to the surface, and then just lie back. You'll float to a safe level. Keep your lungs inflated. Like a beach ball, you'll float on the surface. If you're in more shallow sand, move very slowly backward to firmer ground. Horses, being smarter than humans about such things, will usually stand still and wait to be rescued!

Do you live in an area that freezes in the winter? Ponds and lakes that provide water for your animals in other seasons may be hazardous in winter when critters walk out on ice to get a drink. Rivers pose an even greater threat. If your pet moose walks out on the ice at the shore, he may fall in and be swept away by the current. Spend the time to fence off the danger at the start of winter.

SECTION TWO

SECTION TWO gives a brief overview of the nature of horses and how to handle them. It's very important for any non-horse person who will be in a life-threatening situation to know how your horse is likely to react. What seems obvious to you might never occur to someone not used to dealing with large animals.

This section also addresses emergency horse first aid. **Do NOT expect emergency responders to do any more than what is vital to your horse's survival.** They're not trained in veterinary medicine. A large animal veterinarian should, but probably won't, be called as part of the emergency response team or Incident Command System. You may need to insist on getting this help for your horse. In fact, if you are the person calling in the incident to 9-1-1, make sure you specify the need for medical help for your horse.

Included is a chapter on euthanasia. Most writers choose to glide over this very painful topic with little or no practical information, but there will be times when an animal has suffered a catastrophic injury and needs to be euthanized on-scene.

The thought of euthanizing an animal is heartbreaking for his owner, and can be equally as traumatic for the person performing this service. And, yes, it IS a service to both animal and owner.

Releasing an animal from his pain may be the final act of kindness you and the responder can offer. You will want to make sure it's done as safely as possible, and with the least amount of emotional trauma to you, your horse and others on-scene.

As mentioned in the chapter on euthanasia, equine insurance companies were contacted to find out their policies on emergency euthanasia. While they will honor policies of animals who die in this manner, they have specific requirements and you need to know them beforehand. After the trauma of a trailer accident and the death of your equine partner is not the time to suffer the added shock of finding out that your insurance company won't pay your claim because the accident was not documented properly. "Forewarned is forearmed".

While you're talking to your insurance company, let them know that you are learning about Large Animal Rescue. Some companies are discounting policies for this knowledge.

SECTION THREE

SECTION THREE was written just for you, the horse owner. You'll learn about the mechanics of an emergency call and the structure of the Incident Command System – how the responders will proceed.

Understanding the framework of a response and the language will make your input more credible.

You'll also learn about trailers from the perspective of both your horse and the responders who may be called upon to save your horse.

This section is filled with examples and interesting tidbits of information -- stories of what NOT to do and the consequences; items you can carry to augment those brought by responders; what you need in your first aid kit; and a first aid kit for your rig.

If you would like to build your own "Go Bag" – a collection of useful tools, ropes, and straps – you will find a page listing the contents of a typical bag. Not all emergency responders are going to have every piece of equipment needed for an animal rescue, so having your own to add to theirs will expedite the rescue.

Both you and the responders will find the knot chapter useful. You don't need to know anything about knots. With this chapter you'll be a "knotsman" in no time at all!

APPENDIX

The **APPENDIX** is chock full of information for both you and the rescuers.

In this section is a chapter by Roger Lauze of the Massachusetts SPCA on how to build your own equine ambulance. Roger teaches classes on the subject.

The "Barn Fire Safety" chapter was written with the help of Laurie Loveman, the foremost expert on barn fire safety in North America. It will help make your barn as fire safe as possible.

There is specific information on trailers to aid responders who need to tear into a wreck; and that same information may help you when it comes to choosing the right trailer for your needs.

Rescuers may need to know the pros and cons of using some of their workaday tools around horses, and that can be found under "Tools for Extrication."

Under Resources, "Rescue Equipment" offers tips on adapting everyday items for use in emergencies: how to tie an emergency rope halter; how to fashion a lifting harness; what items to use for rescue straps. This chapter also lists and describes the ultimate in professionally designed and engineered rescue equipment.

The "Resources" chapter also lists LAR instructors. Most will welcome you in their classes, whether you're a horse owner, horse enthusiast, or responder. The more information you have and the better equipped you are, the easier and faster the rescue.

This training is only a few decades old so not many emergency responders are trained in this area, but they need to be. In California, LAR is FSTEP (Fire Service Training and Education Program) certified, which means that the class counts as continuing education for firefighters and may be paid for by the appropriate departments.

You will greatly aid traveling horse owners if you provide your emergency responders and their dispatchers with a list of your local equine veterinarians. Encourage your friends in other areas to do the same. Perhaps this information will save the life of an injured horse, either yours or someone else's.

You may want to include small animal veterinarians since many horse people travel with a canine companion. While you are at it, having a list of willing and available horse haulers might help to clear the scene of an incident more quickly AND save a life!

In the APPENDIX you'll find a sample (MOU) Memorandum of Understanding that your local veterinarians can use to lay out the parameters of the help they are willing to offer to your local emergency responders. Having an MOU in place *before* an incident will make sure everyone is operating by the same rules.

Emergency responders are wary of who will be allowed into such a potentially dangerous situation. Having an agreement beforehand does four things: it provides the veterinarian with access to the scene; it introduces the veterinarian to the emergency responders; it establishes the scope of liability for each; and, it gives the emergency responders a chance to teach the veterinarian about the Incident Command System so everyone on-scene understands the language and knows the chain of command.

An MOU outlines who is responsible for what action and limits the liability if anything goes wrong.

USRider, the equestrian motor plan, has provided us with their excellent forms "To Emergency Responders" and a "Limited Power of Attorney for Animal Health Care" that contain contact information should you be incapacitated. Make copies, share with friends, keep a copy in your "go bag" or at least in the glove box of your hauling vehicle and send a copy to your emergency contact (the ICE number on your cell phone) to have on hand in case you need help

While it's unlikely you'll be asked to be part of the rescue effort of a large animal, if it involves your horse, you are the person legally responsible for his welfare. Make sure he gets the best help possible.

Hopefully, you will learn from ***SAVE YOUR HORSE!*** and will then pass on that knowledge.

Wouldn't you feel safer if you knew that all emergency responders had been trained in Large Animal Rescue and that a copy of this book was part of the equipment carried by horse owners, horse haulers and emergency responders?

Happy trails!

Michelle Staples

This page intentionally blank

SECTION ONE

This page intentionally blank

LETTER TO THE INCIDENT COMMANDER

In this book you will find on-scene aid for emergency responders who have not been trained in Large Animal Rescue (LAR). The book has been critiqued both by experts in the field of LAR and by emergency responders.

The format of the book is simple and efficient. Incidents are sectioned into type: an overturned trailer; a vertical lift; a diagonal drag. Each section is fully self-contained. For instance, if you are removing a recumbent horse from a trailer you need only go to that section to find the steps required and the equipment needed. If your rescue changes from a diagonal to a vertical lift, you can move out of the diagonal section and into the vertical lift section. The steps have been designed to fit easily into the Incident Command System.

If justifying rescue of an animal is a problem, remember this: According to the American Horse Council. "Economic Impact of the Horse Industry in the United States", the economic impact of the horse industry is approximately **$112 billion**.

More than seven million people, in 1.4 million full-times jobs, are involved. This means that the horse industry has a greater economic impact than television and radio broadcasting, and 150% of the railroad transportation industry. It equals the combined output of the tobacco and leather products industries; is 83% of the total of the textile industry and 65% of lumber and wood products. To the federal government, this means an estimated $1.9 billion in taxes each year.

LAW ENFORCEMENT

As in most incidents, law enforcement will take control of the security at the scene, including scene safety and traffic control. Additionally, they may share the legal responsibility of the animals involved with Animal Control, and may be required to euthanize animals.

FIREFIGHTERS

The fire department typically will respond with the technical skills to do the job at hand. Firefighters will be the primary resource for staff, equipment, and radios.

ANIMAL CONTROL

In some areas, Animal Control will be part of your ICS and will make legal decisions about the animals involved, usually being the final authority regarding the animals if the owner is not present or coherent. Animal Control may be responsible for containment, transport, and housing of the animals after the rescue, if the owner is not present to care for her animals.

LARGE ANIMAL VETERINARIAN

The Large Animal veterinarian you call to the scene of the incident will make all medical decisions for the animals, including assessment of their condition, the use of medications, and need to euthanize. He may not be familiar with working in team and rescue situations. The vet may assist in the selection of the most appropriate rescue procedure. Make sure he wears safety gear, especially a helmet.

OWNER

If the owner is on-scene and is capable of making rational decisions, she will have the final say regarding her animals.

Your "RESCUE TEAM" will probably include the following three positions, appointed by the Incident Commander:

HORSE HANDLER

The horse handler interfaces with the owner and veterinarian. He makes the initial inspection of the trailer and the situation, and has initial contact with the horse. In an overturned trailer situation, he checks for people in the living quarters and stalls of the trailer and the presence of propane tanks and cooking appliances in the living quarters, if the trailer is outfitted for living. He advises the IC of the veterinarian's needs, stays by the horse, and monitors his status. He also advises the IC if some action is increasing the risk to rescuers and animals.

EXTRICATION AND PULLING TEAM LEADER

This leader is responsible for set-up and operation of hauling/lifting equipment during extrications. She determines the set up of the haul team and systems, and determines the need for additional help.

CONTAINMENT LEADER

The containment leader organizes people and equipment to capture and contain loose animals. He is responsible for defining a safe containment area. He coordinates handlers if more than one animal is involved and is responsible for control of horses until the responsible party takes over.

YOUR SAFETY

Rescuer safety has been the number one concern from the start. As a horse owner, I know the strength and unpredictability of these animals. I've kept this in mind in writing this book, providing information about the nature of horses as well as the mechanics of the rescue itself.

Knowing a bit about the psychology and physiology of horses will increase rescuer safety. That is why you will find an extensive section on interacting with horses, including everything from how to read body language and how to deal with first aid problems, to the mechanics of everyday contact – how to lead a horse, tie him up, get him up when he's lying down. The experts agree that the *greatest safety issue* is in not understanding how to handle the animals involved. While the section covering interaction is near the back of the book, its importance cannot be overstressed.

You may have had little previous contact with large animals. You may not be accustomed to working with them and it is understandable that you may be uncomfortable about responding to an incident involving them. That is your safety mechanism at work. Not only is this a type of rescue that is not covered in your training, it is downright dangerous in ways you don't understand.

In the resource section of the APPENDIX is a list of people who teach this specialized training. For many of you, continuing education is a job requirement. These classes are an excellent way to learn valuable information while satisfying that requirement. Course material typically covers scene protocol; safety; when to call in a large animal veterinarian; the proper and safe methods to lift and drag large animals; specialized equipment; animal psychology and anatomy; and, how to use what you have on hand to get the job done.

Aside from training information, the APPENDIX lists manufacturers of specialized equipment and, more importantly, how to construct your own equipment, if necessary. Some veterinary hospitals and universities have equipment on hand and are willing to loan it out, but chances are you'll be in an area that makes access to this equipment difficult and time-consuming. This book shows you how to make your own lifting harness, sliding apparatus, and other horse equipment.

Your safety and the success of a rescue may hinge on information about the structure of the trailer involved. This has been addressed in the APPENDIX. You will also find a chart detailing the desirability and pitfalls of using the power tools typically used in rescues.

Not all incidents will involve horses so there is information on other species in the chapter entitled, "Other Livestock". This chapter was written by an expert in the field, with years of hands-on experience, and addresses the issue of how to proceed with an incident involving a commercial livestock trailer.

Another useful section of the book is the chapter on Barn Fire Safety. Collaborating on this section was Laurie Loveman, retired firefighter and horse owner who is the foremost expert on barn fire safety in North America. You'll also find Roger Lauze on building a horse ambulance, Wave Trek Rescue on swift water, and Eric Thompson on ice rescues. Roger teaches an ambulance class through his work with the Massachusetts SPCA, Wave Trek Rescue offers a variety of water classes, and Eric is the founder of the Emergency Equine Rescue Unit in Kansas. See the Resource Section at the end of the book for more on these people.

I hope you'll find this book and the knowledge you learn helpful on-scene, and it will encourage you to become trained in Large Animal Rescue.

RESCUER SAFETY

If the situation is not safe for you and your crew, decline. Your safety is the first consideration!

Keep the horse calm

- Stress is bad for the horse's medical condition, but more important it increases the danger to rescuers
- Horses are afraid of being trapped, loud noises, flashing lights, strange sights and smells, and strangers approaching
- Keep noise to a minimum near the horse
- Have one person approach the horse and to check out the trailer (if one is involved in the incident). All others stay back
- Speak softly to the horse to lower his fear level. Stroke the horse instead of patting him, which is a more aggressive action. Keep your voice a monotone, not "sing-song". Staring is a sign of aggression

Stay away from the horse's head and hooves

Stressed or frightened horses can thrash or strike out at high speed and with great force

Stay out of the danger zone (shown in grey) – directly in front of horse; up to 8 feet to the rear and to the side of rear legs. If the horse is recumbent, beware of thrashing legs and head

- Do not get into water with a horse

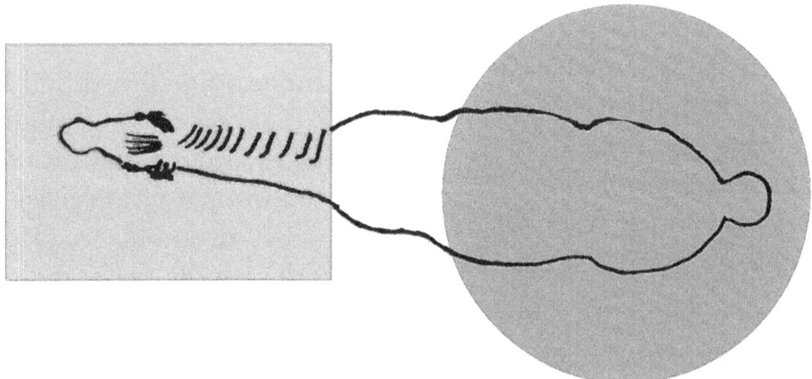

For most situations, do not move the horse without a veterinarian being present

Proper sedation may lessen the danger for the horse and the rescuers. On the other hand, animals are unpredictable in their reaction to sedation and are extremely dangerous when coming out of sedation. The veterinarian may have valuable input to the rescue process.

Consider all horses, sedated or not, as potentially explosive

- You can't hold a panicking horse
- A bolting horse --- even a pony or miniature horse -- can and will drag a person
- All horses are very quick and strong compared with humans
- If your horse is trying to escape, try facing his side and jerking hard on the lead rope to unbalance him. This will only work if you catch the horse quickly before he has "dug in" or he reaches the end of the rope. If the horse is determined to leave, let go

Be aware of your body mechanics

- Horses weigh about 1,000 pounds on average, but can weigh as much as 2,400 pounds
- A horse can be five times or *sixteen times* larger than you
- The reaction time of a horse is the fastest of any domesticated animal

- Establish a safe area for post-extrication
- If the horse gets loose, don't chase him
- Do not try to "rope" the horse like a cowboy; you'll just scare him more
- Do not try to make eye contact with the horse. This is a predatory action and may cause him to run
- Do not move suddenly from side to side in front of the horse or look rapidly from side to side. Keep movement fluid and slow

Use the proper equipment

- Make sure your equipment is strong enough to handle the job. Systems used in a horse rescue must be heavier and stronger than those used for humans. The size, weight and unpredictable behavior of the horse can put an instant, huge load on the system

Contain the horse once he is out

All techniques presented are the safest and the most efficient I could find and have been approved by experts in the field. Consider them your first choice in any situation.

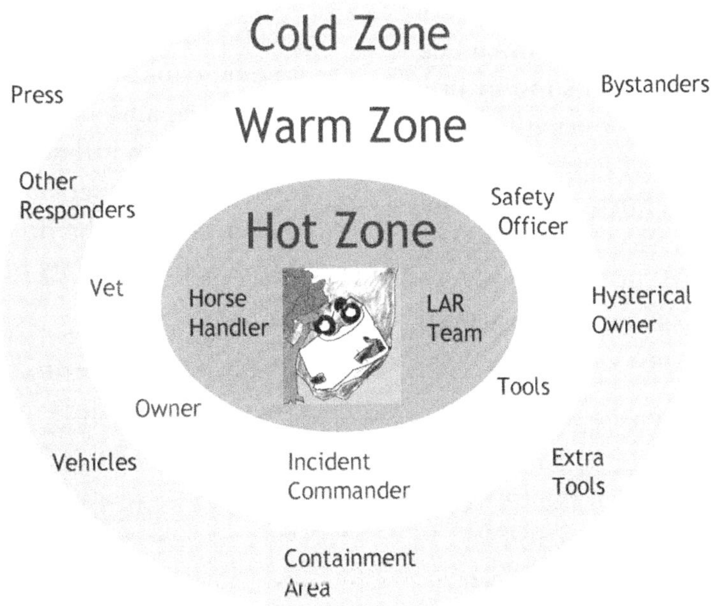

LETTER TO THE VETERINARIAN

Because veterinarian students are not trained in large animal rescue, disaster triage and advanced handling of horses, your first exposure to an emergency horse rescue may be when you are "thrown in at the deep end" following a call from emergency services. This, you will agree, is not an ideal first point of contact. Not only are you not trained in the techniques used to rescue large animals, you are also not trained in safety procedures, crowd control, or how to assign tasks within an Incident Command System.

The Incident Command System described in SECTION THREE is a method used by all emergency responders to identify leadership and resources and ensure the safety of all at the scene.

Teamwork is critical at a rescue scene but your unfamiliarity with the ICS may make you unwilling to accept the word of the Incident Commander, follow his directions, and support his decisions, thus compromising your own safety and that of emergency responders and the horse. Please read the information in SECTION THREE for your own safety and so you can contribute to the smooth execution of the rescue operation.

It can be critical to have a large animal vet on scene for sedation, anesthesia, the welfare of the horse and safety of emergency responders who are probably not familiar with horses and therefore won't understand how medically fragile they are and how explosively they can react.

Your effective, timely action can improve the viability of a horse in distress and have a critical effect on his survival and future usefulness. It is now accepted by vets experienced in LAR that when first treating the horse you should "go big early" with sedation/anesthesia to flood receptors that are already coated with adrenalin.

Accident scenes are usually chaotic, the participants can be emotional and a rescue team that is drawn into the emotion of rescuing a horse can put the team at risk only to find the horse is in such poor condition or injured so badly that it must be euthanized.

Your presence can ensure the safety of all rescuers by making them aware that they must always have an escape route away from the horse in case he panics or reacts adversely to anesthesia, and you can prevent rescuers from inflicting further injury to the victim by ensuring they do not attempt to pull the horse out using ropes tied around its head, neck or legs.

Please wear protective gear – especially a helmet – at all times while at the scene.

Our thanks to MaryAnne Leighton, our Australian partner, for this information

LETTER TO THE HORSE OWNER

For your own safety and that of the emergency responders who are trying to save your horse's life, please:

- Stay calm – your horse can sense your fear and will react to it and, just as importantly, you will find it easier to make rational decisions if you are calm
- Call 911 – do not try to rescue your horse by yourself because you could be seriously injured or killed, you could injure or kill your horse, or you could imperil his life by delaying rescue
- When you call 911, tell the operator you need emergency responders, that a horse is involved, that she should notify a large animal veterinarian if you are away from home or have not already done so, and ask for response vehicles to arrive on scene with their sirens and lights turned off – loud noise will frighten an already stressed horse and could put him into life-threatening shock
- If you are at home or near home, call your own vet if your horse needs to be sedated or have injuries treated
- If you are away from home and your trailer has been damaged, the local animal control may have a vehicle that can carry horses
- **Be patient**
- Do not allow untrained people, no matter how well-meaning, to jeopardise their own safety, your safety and that of your horse by attempting to rescue him
- When emergency responders arrive, allow them to take control of the situation – they are trained, you are not
- One of the biggest challenges to emergency responders at the scene of a horse accident is you, the owner – try to put emotion aside and act rationally if you are asked to leave the immediate area (it will be for a good reason)
- Be respectful of the chain of command of the people who are putting their lives at risk to help you and your horse
- Do not act aggressively towards rescuers and do not try to direct them to perform practices that could endanger their lives – no matter how valuable your horse is or how much you love him, human life always comes first
- If the IC asks for your help during the rescue, make sure you are wearing sturdy boots and a hard hat or helmet
- It is normal for you to feel afraid, angry or guilty – but do not take it out on the rescuers, they are there to help you

Our thanks to MaryAnne Leighton, our Australian partner, for this information

This page intentionally blank

EXTRICATION FROM A TRAILER

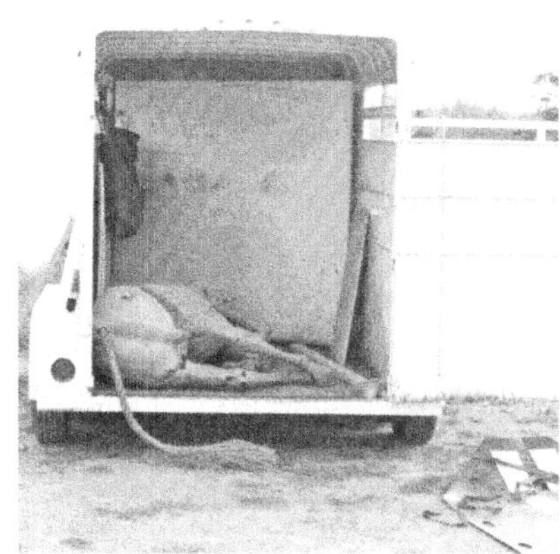

Usually horse trailers, if properly maintained, are very sturdy. As a consequence, most horses are not severely injured in a rollover, and, unless the trailer itself is a hazard, the horse is usually not at risk by staying in the trailer until the veterinarian arrives. The exceptions are: where more than one horse is occupying the trailer and the horses are lying on top of one another, usually caused by the collapse of flimsy dividers; the horse's legs have gone through openings in the trailer (windows, through bars); and where the horses are tied inside to the side of the trailer. In the latter case, the horses could be lying on their heads or could be hanging by their tie straps and strangling.

Always use the safest, simplest, and least intrusive method of removing the horse.

- Self extrication
- Assisted self extrication
- Drag or roll the horse out
- Lift the horse out
- Use power tools to cut the horse out

Where to get help:

- Animal Control and Law Enforcement
- Local horse rescues and horse councils
- CART (County Animal Response Team) SART (State Animal Response Team), and DART (Disaster Animal Response Team)
- Large animal veterinarians

Contain horses using the Three Fs: Fencing; Friends (they're herd animals); Food

A large animal in entrapment is a HazMat incident:
A material or substance that poses a danger to life, property, or the environment if improperly stored, shipped, or handled.

STEP ONE:

SCENE SAFETY As with any incident, scene safety is your first concern. Scene safety is different in intensity and scope when you factor in large animals. Check for down power lines, leaking fuel or other hazmat problems, batteries in and around the trailer, traffic and human interference, loose animals, and the overall stability of the vehicles involved. You may want to unplug and unhitch the trailer from the tow vehicle and chock the tires.

SECURE THE AREA Make sure the area around the trailer or accident site is safe for people to approach. Have only one person approach the trailer to check the site. Approach quietly and slowly so as not to frighten the horse. When you look into the trailer, use the smallest possible opening. Do not open any doors or windows, and cover all openings with tarps to discourage the horse from trying to escape. A thousand pound horse WILL try to exit through a two foot square window. In hot weather, keep the horse cool with fans or spray with water.

Rescuer safety is your first concern. Keep all non-responders out of the area. Never go into the trailer with the horse.

STEP TWO:

CALL THE VETERINARIAN -- a large animal veterinarian who is experienced with horses. There could be more than an hour wait time before a veterinarian can be on-scene. You will need a veterinarian to sedate, and later treat, the horse. The veterinarian can also determine if the horse is so grievously injured he can't be saved, and will have the drugs to humanely euthanize the horse. Having a veterinarian in attendance will decrease the danger to rescuers, as well – a major consideration. If your area mandates it, inform Animal Control of the incident. Please share this book with them.

STEP THREE:

APPOINT A HORSE HANDLER to stay with the horse for the duration of the rescue, talking softly and calmly. This should be the person with the most experience around horses. The horse handler will monitor the horse's physical and emotional condition, interact with the veterinarian, as well as watch for obstacles when moving the horse.

Monitor the horse's body language, such as ear direction, to follow the horse's focus. If the owner is calm enough you might need to use her. Keep in mind that the owner's first concern is the horse, not YOUR safety! Using the owner could present a liability issue; if this person is your horse handler you may want to station her in the warm zone with the Incident Commander and the Veterinarian.

Ask the owner the following:

- Has she been trained in Large Animal Rescue techniques?
- Is she willing to observe her horse from a distance, passing along information about the horse's state of mind?
- Has the horse been sedated (pass along this information to the veterinarian)?
- Are there any known issues with the horse that will further jeopardize your safety or his? (Hates men; can't tie; likes to kick)

As much as possible, keep other people away from the trailer. If more than one horse, provide a handler for each.

Find out if the involved horse is insured. There may be existing protocol about handling him. As you await the arrival of the veterinarian, and if it is possible and safe to do so, take and record the horse's vital signs. (See SECTION TWO: "Horse First Aid for Emergency Responders"). Use the wait time to prepare the scene, setting up a safe containment area for the rescued horse.

Once the horse is out, the handler will be handling the horse with a lead rope. It is called a lead rope because she will lead the horse with it. **SAFETY NOTE TO HANDLER:** When leading the horse: DO NOT WRAP THE ROPE AROUND YOUR HAND! Standing on the left side of the horse and facing forward, hold the rope under the horse's chin with your right hand, making sure the rest of the rope is not looped, but fold it back and forth and hold the folds in your other hand.

NOTE: **Use the lead rope for leading/directing/containment and to keep the horse from slamming his head on the ground. A horse needs free movement of his head when he is trying to stand up, to walk, and to balance. Don't try to pull him – the steady pressure on his head will not allow him to move in your direction. Allow him to help you in his rescue.**

KEEP THE HORSE CALM This will be your main focus, throughout the rescue effort. Keep the area free of chaos.

- Horses defend themselves by fight or flight
- When they feel trapped, their stress level shoots up. This can be disastrous for their medical condition and makes the danger much greater for rescuers. It is difficult to sedate a stressed horse, it may take up to 8 times the normal dose of sedative, so it is better to keep the horse calm in the first place
- Panicked horses may thrash and strike out with hooves and teeth at anyone who gets close, slam with their heads or hindquarters, or run over or roll over anyone in their way. The calmer the horse, the greater the safety for emergency responders
- Gently stroke– don't pat – the horse (less aggressive). Avoid eye contact or sudden movement. Speak softly in a monotone voice
- Stallions, and mares with foals, are especially dangerous

AVOID THE DANGER ZONE! – directly in front of the horse; up to 8 feet to the rear and to the side of rear legs, depending on the size of the horse.

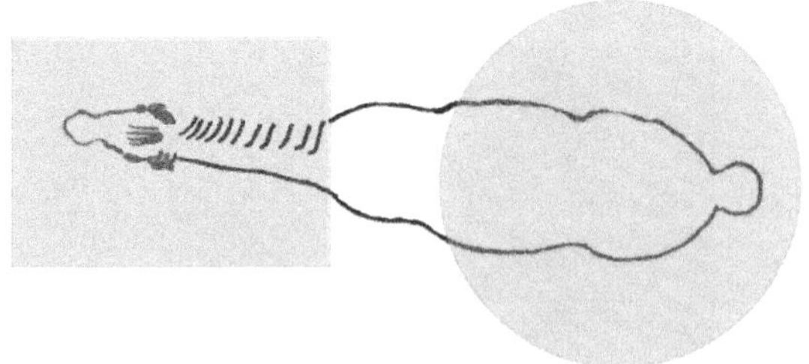

KEEP THE NOISE LEVEL DOWN Extreme light/dark contrasts such as those caused by flood lights used at night are frightening to a horse because he will not be able to see properly. By keeping the area free from chaos and loud noises you will lessen his urgency to flee, thereby lessening the risk to rescuers.

If you use a generator to power your tools, move it as far from the trailer as possible.

DO NOT OPEN THE DOOR OR WINDOWS A horse who feels trapped or frightened will try to escape, even through a very small opening, causing injury. Walk around the trailer to assess the situation before attempting to gain access to the horse. Once you have done this, determine the location of the horse's head and quietly and slowly open the closest, smallest opening, then close it immediately. Keep openings secure until the last minute!

CHECK THE TRAILER FOR HUMANS Although it is not legal for a person to ride in a trailer, you may find one in with the horses. If the trailer has a living unit at the front, look for people and pets there as well.

STEP FOUR:

GATHER THE EQUIPMENT NEEDED to perform this type of rescue.

- Rescue Strap – either commercial, or made from 4 inch webbing or 2 ½ to 4 inch fire hose 30-50 feet long, two ply tow strap with flat loops, 4WD Recovery Kit (see APPENDIX: "Chart for Hose Dimensions")
- Soft padding material to place under horse's head and over his eyes for protection, and protection from cutting tools
- A pike pole/ceiling hook, shepherd's crook, or other hooked pole – wrap the sharp points
- Rescue rope and 2 inch webbing, at least 30 feet long
- If you will be dragging the horse away from the trailer you will need something to drag him on -- either a commercial Rescue Glide, Rescue Mat, inflatable Rescue Path, or something solid and non-shredding like plywood or heavy tarp
- Flood lights (the rescue may take awhile!)

It is a good idea to make a pair of earplugs for the horse. Muting the sounds associated with the rescue may help calm him. Stuff 6 or 8 cotton balls in each foot of a pair of cut off pantyhose or thin socks and tie a knot at the open end. This knot gives you a handhold for easy removal. The smell of turnout gear – smoke, diesel, gasoline, blood – can cause uneasiness as well.

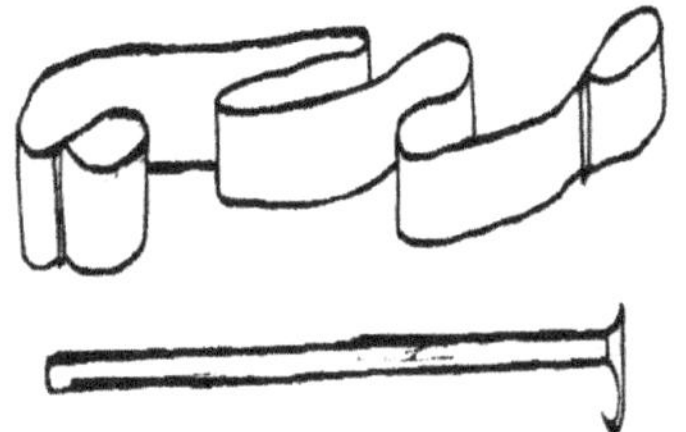

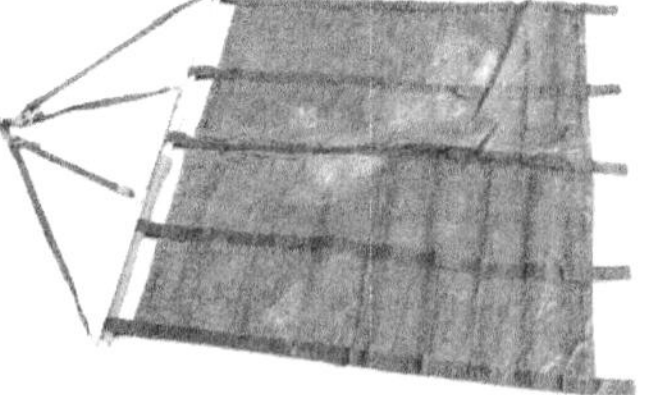

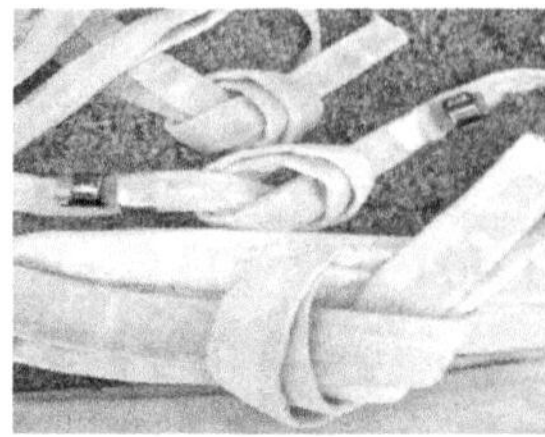

Rescue Strap, pike pole, Rescue Mat and hose
Mat photo courtesy of Timothy Collins; Hose photo courtesy of Vicki Schmidt

> *From Vicki Schmidt, Maine State Fire Instructor:* "Fire attack line, usually 1 ½- 2 ½ inch, can be used. Fifty foot sections of "out of service" or retired hose line with the couplings removed are easier to handle than hose that is still in service. One hundred foot sections can also be used. If preparing hose lines for LAR, a variety of both sizes is recommended. Tie the ends of the hose in a water knot (see SECTION III: "Knots for Horse Owners"), and secure the loose ends with duct tape if possible. This is especially handy if the emergency requires the use of in-service hose."

Ask the owner/hauler if she has any rescue equipment in her vehicle or the trailer. She may have a "go bag" with some of the equipment needed for a rescue.

STEP FIVE:

CREATE A SAFE AREA BEHIND THE TRAILER for the horse once he is removed from the trailer. The best equipment to use is portable panels. Many horse people carry them for stopovers and shows. Five or six foot plastic construction fencing reinforced with rigid PVC pipe and held by rescuers can also be used to contain a horse. Another possibility is to park vehicles, nose to tail, around the opening to the trailer. Do NOT use vehicles that are low to the ground, such as sports cars. A panicked horse can easily jump an obstacle 5 feet high. If the horse is frightened once he is free he may run out into traffic or into new danger.

STEP SIX:

IF THE HORSE IS WEARING A HALTER, have the handler attach a strong, soft ½ inch or ¾ inch rope (Rescue Rope, Kernmantle rope, parachute cord), at least 20-30 feet long. The rope should be long enough to reach from the horse to the handler who is on safe ground outside the trailer at the point of exit. This is called the "lead rope". If the horse isn't wearing a halter, and none is available, use this rope to make an emergency halter. (See APPENDIX: "Rescue Equipment" for directions). Do not use a bridle, even if it is already on the horse. This will result in damage to the horse, and possible damage to the rescuers.

There is also a security issue; the straps of a bridle are usually made of leather which stretches and are easily broken. Horses led by bridles will often throw their heads around to avoid pain from the bit – a piece of metal – in their mouths.

Halter vs. Bridles

STEP SEVEN:

IF THE HORSE IS STANDING, UNINJURED, AND AN EXIT IS AVAILABLE, lead or back him out into the safe, enclosed space you have created. Assume that the horse will panic, and have everyone out of the area except for the handler. As the horse calms down, take him to a quieter area or rescue transport.

IF THE HORSE IS RECUMBENT (lying down), UNINJURED and CALM, and the horse's life is in danger and he needs to be moved immediately (trailer on fire or falling over a cliff), it may be possible to extricate him without sedation, and using his tail. Braid the rope into the tail hair **BELOW the tailbone** (feels solid coming out of the spine and is about a foot long) using a Sheet Bend/Becket knot (See SECTION III, "Knots for Horse Owners").

The pull needs to be in a straight line with the spine. DO NOT ATTACH TO A VEHICLE. Pulling needs to be slow, fluid, totally controlled, and by hand only. If the horse gets caught on something or friction becomes too much, stop pulling or you will cause severe injury.

DO NOT ATTACH ANYTHING TO THE TAILBONE.
Below is a picture of a horse whose tail was used to pull her out of a trailer. The rope was hooked up to a tractor. The horse died from her injuries. **This maneuver is definitely NOT recommended!**

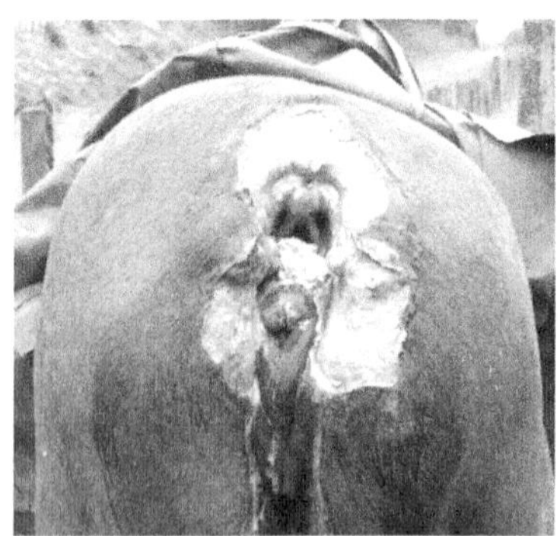

Photo courtesy of Allan Schwartz

STEP EIGHT:

IF THE HORSE IS INJURED, or the EXIT IS BLOCKED, do not do anything more with the horse until the veterinarian arrives and sedates him. Otherwise the horse will remain frantic and is a much greater danger to himself and the rescuers. Be aware, however, that horses can react unpredictably to sedation.

A HORSE CAN APPEAR TO BE ASLEEP ONE SECOND AND EXPLODE THE NEXT, EVEN WHEN SEDATED. This cannot be overemphasized. A horse can go from standing to a complete connecting kick in about **1/3** of a second.

STEP NINE:

CREATE A RESCUE OPENING if the horse can't be removed through an existing door. If you have to use power tools (See APPENDIX: "Use of Power Tools in Extrications"), start them some distance away, and gradually bring them closer to the trailer, to avoid panicking the horse. Cover the horse or otherwise shield him, especially if you are using a tool that creates sparks. Also be aware that horse trailers typically contain flammable materials such as wood shavings and hay. Horses are obligate nose breathers which means they can ONLY breathe through their noses. **Do not block the horse's nose**.

STEP TEN:

IF THE HORSE IS RECUMBENT you will have to pull him out. **DO NOT PULL HIM OUT BY HIS LEGS, TAIL, NECK, OR HEAD**. All can cause severe damage to the horse.

Use your rescue strap from Step 4. The straps will pull on strong bony structures of the horse, avoiding further injury.

Before extrication, the horse's lead rope (attached to the halter) will be pulled out through the exit opening so that it can be used to control the horse once he is free. Be sure to protect his head.

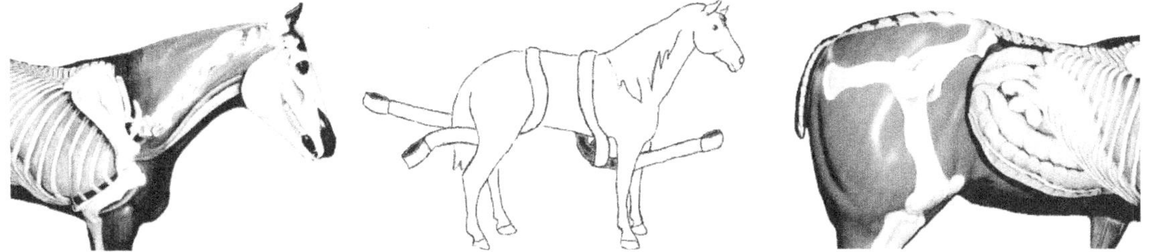

Backward/Forward (Lark's Foot shown) straps use the bony structure of the horse

STEP ELEVEN:

WHEN APPLYING THE STRAP you must AT ALL TIMES stay out of the "line of fire" of both the horse's hooves and his head. Use a pike pole (sometimes called a ceiling hook or closet hook) or shepherd's crook to grab or push the strap around the horse. Wrap sharp points of the pole and other equipment to protect the horse.

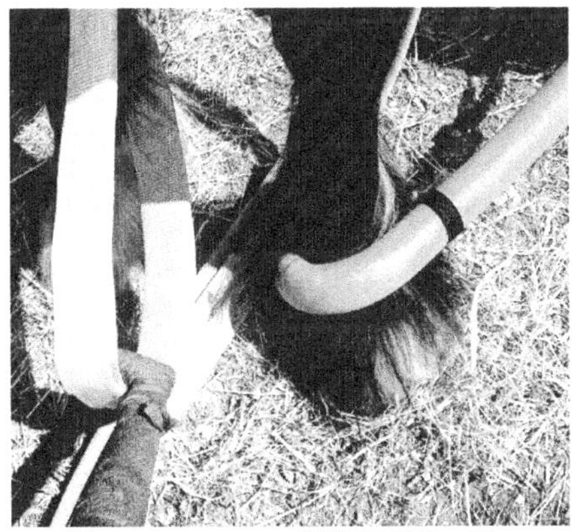

Photo courtesy of Vicki Schmidt

Start by using a length (approximately 20 plus feet) of 1½ to 2 inch webbing. Objective: Get the webbing under the horse's body and use it to pull the rescue strap through after it. If there is access at the front of the trailer, or the side opposite the exit opening, use this. If not, use the opening you created.

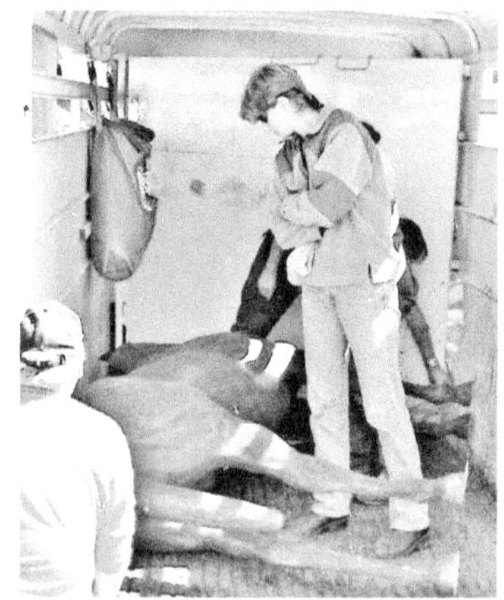

Safe versus **not** safe! Wear a helmet if one is available.

Rescuer 1: (at the end away from the exit) will push most of the webbing into the area beside the horse using the pike pole, keeping at least six feet with which to pull.

Rescuer 2: (at the exit or rescue opening) take the pike pole and go to the door or rescue opening, USING THE STRUCTURE OF THE TRAILER AS A SHIELD. If there is nothing to use as a shield, stay out of the DANGER ZONE.

The webbing, at corner A, is pushed across to corner B. Rescuer 2 pulls it along the side to his position at corner C.

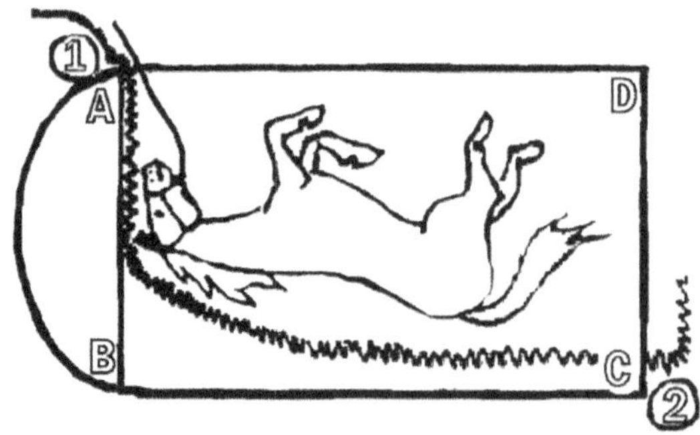

The webbing is then worked under the horse. Rescuers 1 and 2 alternately pull on the webbing in a "sawing" motion. This will result in the webbing lying under the horse's neck. There is a natural hollow under the neck at the shoulder and under the body in front of the back legs.

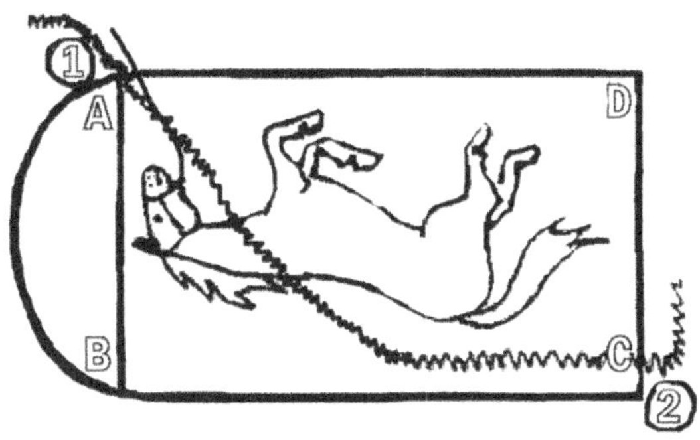

The rescue strap is attached to Rescuer 1's end of the webbing, and it is pulled through until it is lying under the horse's neck and along his back. Remove the webbing. If the veterinarian is on-scene he will probably have lubricant that will make it easier to slide the strap under the horse.

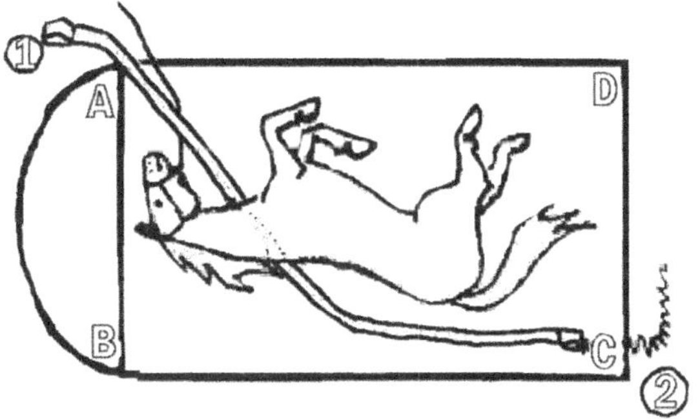

Rescuer 2: Use the pike pole to reposition the strap, so it runs under the horse at a right angle to his spine. This may take a few helpers on each end, as it will not be easy to pull this under the horse!

At this point the rescue is different for forward and backward extrications.

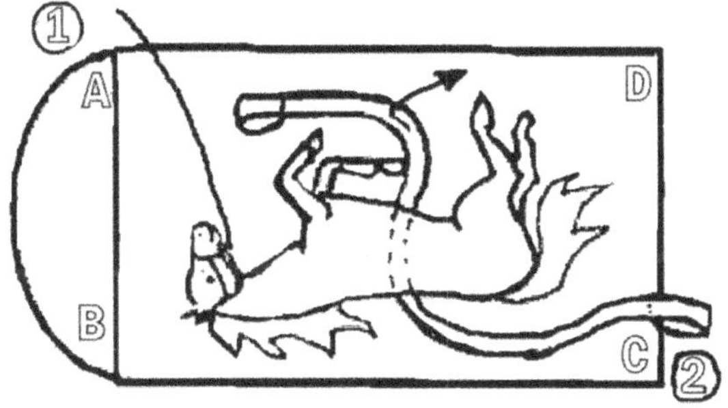

BACKWARD EXTRICATION The rescue strap should be positioned under the horse's "waist," just in front of his hind legs and hips. (The hips are where a human's would be if he were on his hands and knees.) Using the pike pole, the end underneath the horse is pulled backwards through the hind legs. The top end is draped over the horse's waist and pulled back through the hind legs as well. The two ends are secured to the pulley system or taken up by the pulling team, and the horse is ready to be pulled out. Adjust the angle of drag so the strap does not slide off the horse as the pressure of the strap on the pelvis may cause the hips to spread. Do not tie the legs together in an effort to avoid slippage as this will place a dangerous strain on the legs and hip joints.

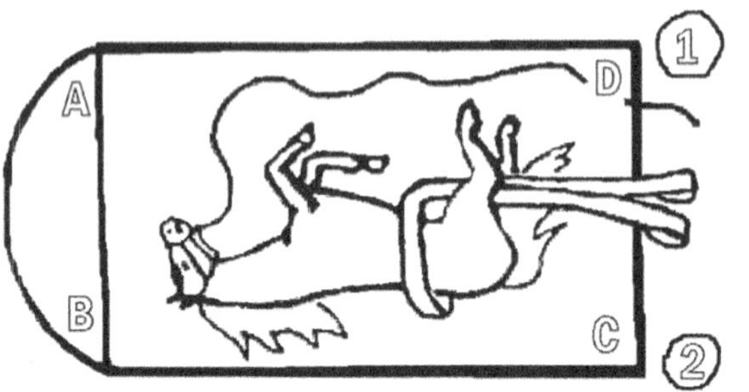

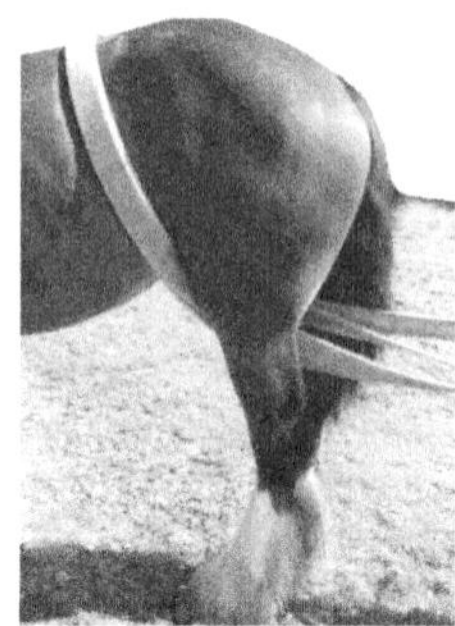

Photo courtesy of Vicki Schmidt

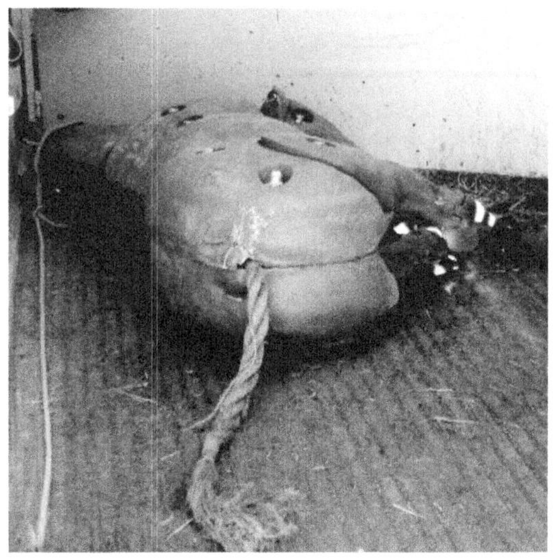

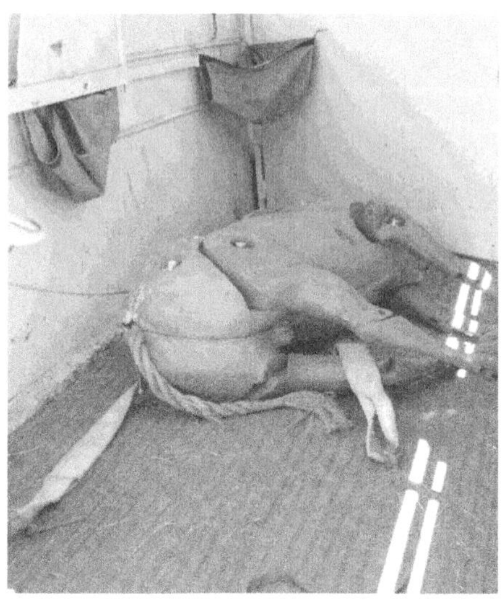

If the horse has to be pulled out over an obstruction, make a ramp with boards or some other strong material like plywood. Wedge one end under the leading end of the horse, and prop the other on the obstruction, so the horse can slide out over it. Since the stability of the ramp can be problematic and the process of getting the ramp in under the horse can be dangerous, try to remove the obstruction first.

FORWARD EXTRICATION The rescue strap should be under the horse, right behind the front legs. The end of the strap on the bottom is lifted and pulled through the front legs with the pike pole and passed to Rescuer 1.

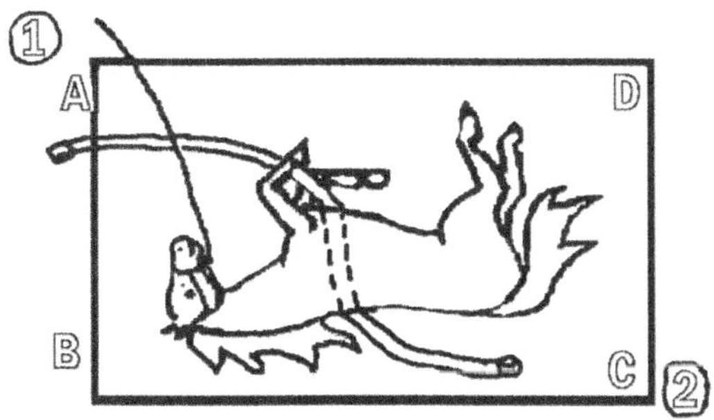

The other end of the strap, which has been freed from the webbing, is lifted over the horse's back, and then also pulled through the horse's front legs. Less secure: The two ends are attached to your pulley system. More secure: the Lark's Foot (picture below): The longer end of the strap is pulled through the loop (or slit) at the other end of the strap, so that there is a "noose" around the horse's middle. Attach the long end to the pulley system. The Lark's Foot puts significant pressure on the ribs/sternum, but can be considered where more security is needed.

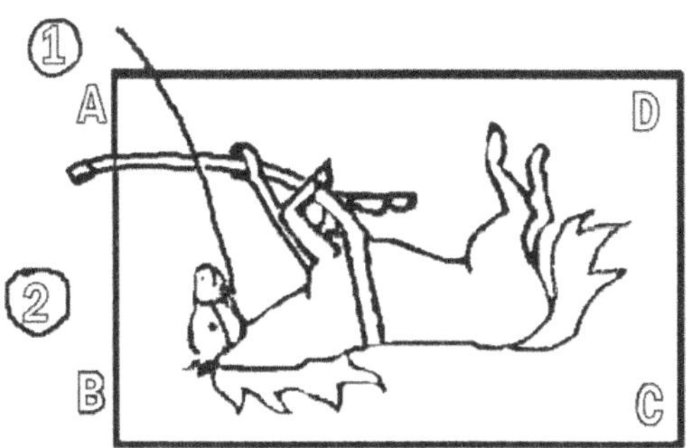

STEP TWELVE:

PULLING THE HORSE OUT To avoid panicking the horse, *CREATE QUIET AND CALM IN THE RESCUE AREA.* If possible, turn off all motors, power tools and emergency lights. Keep all radios out of the immediate vicinity. All you should hear are the I.C. and the horse handler.

Horses may weigh over 2,000 lbs. You should have a good rope system, such as the one shown here.

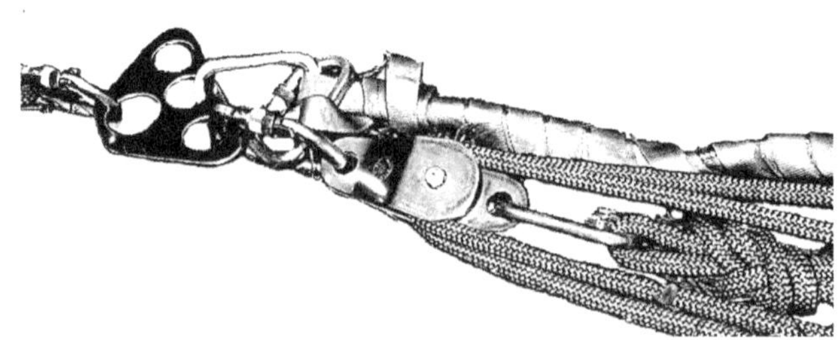

Even with a rope system you will need a lot of people for your haul team. Consider using bystanders. **Never use a vehicle to pull unless there is ABSOLUTELY no other choice**. If the pulling has to stop for a caught leg or a tangled strap, vehicles take much longer to respond. This can be disastrous for both horse and rescuers.

"HORSE HANDLER" CALLS THE SHOTS Until now, the Incident Commander has been in charge. But once the horse is ready to be moved the horse handler becomes the one who calls the shots. The handler watches the horse at all times, and decides when to pull or stop pulling, depending solely upon the condition of the horse.

The handler signals the pulling to begin once all rescuers are in their places, all straps in the right position, padding is in place to protect the horse's head, and the horse is as calm as the handler and chemicals can make him.

The instant any problem develops, the handler immediately gives the signal to STOP. This can be for a caught leg, a strap working loose, the horse panicking, or anything that may adversely affect the horse or the handler.

ONCE THE HORSE IS READY TO BE PULLED, the handler moves to the door (or the rescue opening), and using the pike pole, brings the lead rope out through the exit. This way, the handler will have control over the horse once he is freed. DO NOT USE THE LEAD ROPE TO PULL THE HORSE.

PULL SLOWLY AND STEADILY Again, keep the noise and confusion down. With the use of the pulley, you may be able to locate the haul team off to one side. All members of your pulling team should be on one side of the haul rope so that if the horse panics they can drop the rope and run in one direction.

STEP THIRTEEN:

THE HORSE IS OUT At this time, it is CRITICAL to avoid any unnecessary noise or confusion. You want the horse to remain calm. Even after sedation, some horses can "wake up" in an instant, injuring themselves and any rescuers around them.

A panicked horse can "explode" in a fraction of a second. This is part of his survival mechanism. He may rear up, kick, or run blindly in any direction, including over top of anyone in his way. He may not see a person in his way if he is panicked.

A horse coming out of sedation can injure himself and those around him trying to stand on wobbly legs or even falling while still under the influence of the sedatives. Remember to have your veterinarian on-scene to sedate the horse and use panels or temporary fencing to prevent escaping horses. Allow the horse to lie calmly; do not rush him to get up.

RIGHTING AN OVERTURNED TRAILER

Occasionally, in a trailer rollover, the integrity of the trailer has not been seriously compromised, and the horse is uninjured and in a "standing" position inside the trailer. It is also possible that the only method available to remove the horse could be to right the trailer and let the horse walk out. This is a judgment call, usually made in conference between the veterinarian and the rescuers. There are a few criteria that need to be satisfied when considering this method.

Is the integrity of the trailer such that it will withstand the procedure?

Will the horse be standing when the trailer is righted?

Are the injuries sufficiently minor to withstand the procedure? This is a decision for the veterinarian on-scene.

Do you have the manpower and equipment necessary to proceed?

This is a controversial procedure. Advocates say there are advantages to using this method. The first and foremost is rescuer safety. Even with the possible inoperability of the trailer, the horse IS contained inside. He can be calmed, then backed or led out when it is safe to do so. There is also less possibility of injury to the horse from collapsing trailer parts, sparks from power tools, or his own panic. And, if there are many horses in the trailer this method could be an efficient use of rescuers' time and cause less stress on the horses.

Dr. Kathleen Becker of Kentucky states, "*I can visualize a situation where a trailer was caught in a ravine, on its side or upside down with all typical cutable exits blocked or too dangerous to utilize, that might require such a controlled roll to make access. And, while the technique is taught for a trailer on its side, the principles are the same in almost any orientation. And while, yes, it would be a technique that would not be employed very often, I do believe that it's teaching IS valuable, even if the goal is to STABILIZE the trailer without completely rolling 90-180 degrees.*

The exercise forces the rescuer to analyze the few good anchors on the trailer and use those points to their advantage. Any information is a plus. This is just another tool in the rescuer's tool box of knowledge and techniques to draw upon in a situation."

Others feel it is unnecessary and even harmful.

Code 3 Associates in Colorado says, "*We have never seen the need for uprighting a trailer with livestock or animals still inside. The number of instances where this procedure would be needed is so extremely limited and minimal that it would be irrelevant to teach it as a curriculum subject. It's our view that this causes more injuries and increases stress on the animals. We teach removal of the animals prior to uprighting a trailer, which we leave to the fire department and/or tow trucks, who have the proper equipment and expertise to do this. Although some rescuers do advocate this method of uprighting the trailer first, we do not teach it and we do not recommend it."*

Jennifer Woods of Alberta, Canada's J Woods Livestock Services and Coordinator of the Alberta Livestock Incident Response Plan, weighs in, as well.

"I have spent eight years in the field responding to motor vehicle accidents involving livestock, and under no condition would I ever upright a loaded trailer, nor do I teach people how to do it.

First, there is no way you can do it without further injuring the animal(s). The animal will begin to struggle and get up as soon as the trailer starts to upright and that is very dangerous, especially if there is more than one animal in the trailer. Second, there is no way for you to evaluate the true structural integrity of the trailer. Trailers may begin to fall apart during the uprighting process, especially with the weight of the animals in the trailer, further injuring the horse and putting the safety of responders in jeopardy.

I have seen dead animals fall out of trailers during uprighting. I have documented, investigated or responded to over 200 accidents and I have never had a situation where anyone other than the tow truck company uprighted the trailers following extrication of all live animals. They are trained professionals who have the skills and equipment to perform that job and have all the liability issues covered.

If you choose to right an overturned trailer before removing the animals, take into account the following.

Be sure all personnel know the plan. Have a "spotter" – a person watching the operation – who will monitor the operation and advise the "pullers" regarding any problems.

Each horse will need a "handler". Attach a halter and rope to each. The rope should be long enough to reach from the horse to the handler who is standing clear of the trailer and on safe ground. Do not hold the rope while the trailer is being uprighted.

Attach your ropes or straps to secure points on the trailer such as the axle or upright struts.

Run one strap from the "low side" of the trailer, over the top, and attach to a rope system

Run another strap in the opposite direction from the "high side" of the trailer. It will also run over the top and attach to a rope system.

Fix the "low side" rope system to a secure anchor. The axle of a vehicle or a large tree would be secure enough. Have personnel on the live end of this system to let the trailer down onto its wheels in a slow, controlled fashion.

The ropes running from the "high side" of the trailer will also be manned with personnel who will be responsible for uprighting the trailer.

Attach a rope to the tongue at the front of the trailer. This rope will keep the trailer from skewing as the trailer is being uprighted, so it should be securely anchored as well. When considering the rope system you will use, take into account the weight of the trailer plus the weight of the animals and the equipment. Typical weights of trailers can be found in the APPENDIX: "Trailer Information".

Dr. Tomas Gimenez discusses trailer integrity

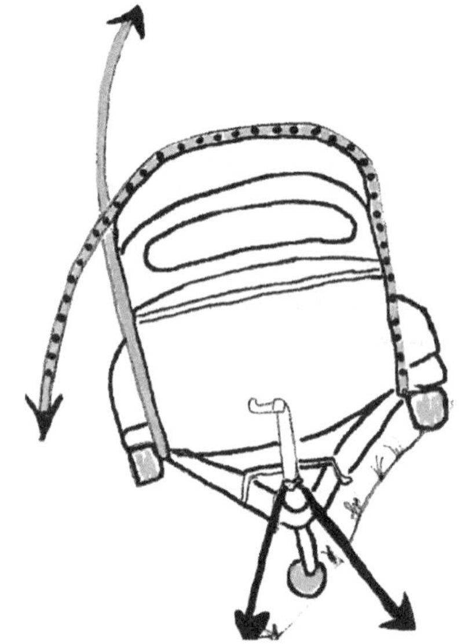

VERTICAL EXTRICATION

Photo Courtesy of Dr. Rebecca Gimenez

Like many prey animals, horses become very uncomfortable when they are lifted off the ground. What you will often find is that a horse, while in the air, becomes passive, hanging like the horse in the picture above. The danger comes when the horse is first raised off the ground, and again when he is set back down on the ground. He may kick, seem to explode in every direction at once, or try to run. Be sure to keep your personnel far away from the horse's space and use a quick release hauling system.

Always use the safest, simplest, and least intrusive method of removing the horse.

- Self extrication
- Assisted self extrication
- Drag the horse out
- Lift the horse out
- Use power tools to cut the horse out

Where to get help:

- Animal Control and Law Enforcement
- Local horse rescues and horse councils
- CART (County Animal Response Team) SART (State Animal Response Team), and DART (Disaster Animal Response Team)
- Large animal veterinarians

Contain horses using the Three Fs: Fencing; Friends (they're herd animals); Food

A large animal in entrapment is a HazMat incident:
A material or substance that poses a danger to life, property, or the environment if improperly stored, shipped, or handled.

STEP ONE:

SCENE SAFETY As with any incident, scene safety is your first concern. Scene safety is different in intensity and scope when you factor in large animals. Check for down power lines, leaking fuel and other hazmat issues, unstable ground and trees, wildlife such as snakes, human interference, and stable access to the area for your vehicles.

SECURE THE AREA Make sure the area around the site is safe for people to approach. Have only one person approach the horse to check the feasibility of rescuing. Approach quietly and slowly so as not to frighten the horse. If the incident involves a ravine, cliff edge or mine shaft, make sure the ground is solid before approaching. If the incident involves a well, septic tank or "black water", check the air quality and decontaminate the horse once he's out. If there is any doubt about the integrity of the ground, personnel approaching the edge should wear a rescue line and harness, approach from a different direction, or decline to conduct the rescue.

Rescuer safety is your first concern. Keep all non-responders out of the area.

STEP TWO:

CALL THE VETERINARIAN -- a large animal veterinarian who is experienced with horses. There could be more than an hour wait time before a veterinarian can be on-scene. You will need a veterinarian to sedate, and later treat, the horse. The veterinarian can also determine if the horse is so grievously injured he can't be saved, and will have the drugs to humanely euthanize the horse. Having a veterinarian in attendance will decrease the danger to rescuers, as well – a major consideration. If your area mandates it, inform Animal Control of the incident. Please share this book with them.

STEP THREE:

APPOINT A HORSE HANDLER to stay with the horse for the duration of the rescue, talking softly and calmly. This should be the person with the most experience around horses. The horse handler will monitor the horse's physical and emotional condition, interact with the veterinarian, as well as watch for obstacles when moving the horse. Monitor the horse's body language, such as ear direction, to follow the horse's focus. If the owner is calm enough, you might need to use her. Keep in mind that the owner's first concern is the horse, not YOUR safety! Using the owner could be a liability issue; if this person is your horse handler you may want to station her in the warm zone with the Incident Commander and the Veterinarian.

Ask the owner the following:

- Has she been trained in Large Animal Rescue techniques?
- Is she willing to observe her horse from a distance, passing along information about the horse's state of mind?
- Has the horse been sedated (pass along this information to the veterinarian)?
- Are there any known issues with the horse that will further jeopardize your safety? (Hates men; can't tie; likes to kick)

As much as possible, keep other people away from the area. If more than one horse, provide a handler for each.

Find out if the involved horse is insured. There may be existing protocol about handling him. As you await the arrival of the veterinarian, and if it is possible and safe to do so, take and record the horse's vital signs. (See SECTION TWO: "Horse First Aid for Emergency Responders"). Use the wait time to prepare the scene, setting up a safe containment area for the rescued horse.

Once the horse is out, you will be handling him with a lead rope. It is called a lead rope because you will lead the horse with it. **SAFETY NOTE TO HANDLER:** When leading the horse, DO NOT WRAP THE ROPE AROUND YOUR HAND! Standing on the left side of the horse and facing forward, hold the rope under the horse's chin with your right hand, making sure the rest of the rope is not looped, but fold it back and forth and hold the folds in your other hand.

NOTE: **Use the lead rope for leading/directing/containment and to keep the horse from slamming his head on the ground. A horse needs free movement of his head when he is trying to stand up, to walk, and to balance. Don't try to pull him – the steady pressure on his head will not allow him to move in your direction. Allow him to help you in his rescue.**

KEEP THE HORSE CALM This will be your main focus, throughout the rescue effort. Keep the area free of chaos.

- Horses defend themselves by fight or flight
- When they feel trapped, their stress level shoots up. This can be disastrous for their medical condition and makes the danger much greater for rescuers. It is difficult to sedate a stressed horse, it may take up to 8 times the normal dose of sedative, so it is better to keep the horse calm in the first place
- Panicked horses may thrash and strike out with hooves and teeth at anyone who gets close, slam with their heads or hindquarters, or run over or roll over anyone in their way. The calmer the horse, the greater the safety for emergency responders
- Gently stroke – don't pat – the horse (less aggressive). Avoid eye contact or sudden movement. Speak softly in a monotone voice.
- Stallions, and mares with foals, are especially dangerous

AVOID THE DANGER ZONE! – directly in front of the horse; up to 8 feet to the rear and to the side of rear legs, depending on the size of the horse.

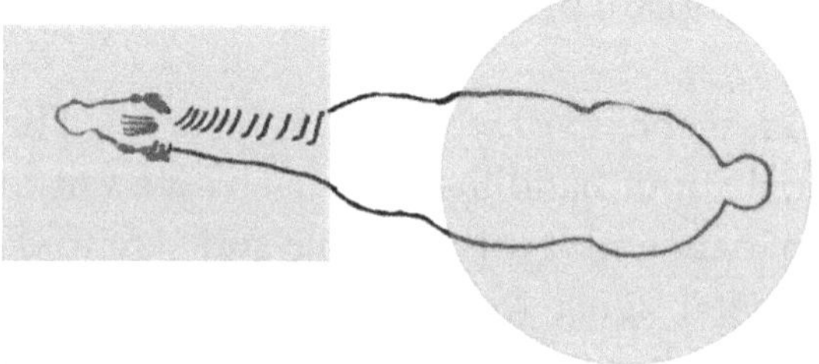

KEEP THE NOISE LEVEL DOWN Extreme light/dark contrasts such as those caused by flood lights used at night are frightening to a horse because he will not be able to see properly. By keeping the area free from chaos and loud noises you will lessen his urgency to flee, thereby lessening the risk to rescuers.

If you use generators to power your tools, move them as far from the rescue site as possible.

STEP FOUR:

GATHER THE EQUIPMENT NEEDED to perform this type of rescue.

- Vertical Harness – either a commercial "Two Point" harness, or one made from 2 ½ to 4 inch fire hose, 4-6 inch webbing, or 2 inch tubular web, or a "Figure Eight" harness made from wide, soft cotton rope. (see STEP NINE: "Apply the Vertical Lift Harness" to determine the type of harness to use, plus APPENDIX: "Chart for Hose Dimensions", and APPENDIX: "Rescue Equipment: Using What You Have" for directions. See STEP EIGHT: "Establish a Means of Pulling the Horse up Vertically" for confined space vertical lifts)

- Rope to secure the horse into position
- Soft padding material to protect the horse from the straps
- A mechanism to raise the horse
- A pike pole/ceiling hook, shepherd's crook or other hooked pole – wrap the sharp points
- If dragging the horse is a possibility you will also need to get something to drag him on --- either a Glide Mat, Rescue Mat, inflatable Rescue Path, or something solid and non-shredding like a heavy tarp or plywood. (See STEP NINE: "If the horse is injured")
- Flood lights (the rescue may take awhile!)

An open water rescue might use floating devices, and contained water rescues such as from swimming pools may need floats. When lifting is a possibility, get your equipment ready beforehand. (See "SPECIALTY RESCUES")

It is a good idea to make a pair of earplugs for the horse. Muting the sounds associated with the rescue may help calm him. Stuff 6 or 8 cotton balls in each foot of a pair of cut off pantyhose or thin socks and tie a knot at the open end. This knot gives you a handhold for easy removal. The smell of turnout gear – smoke, diesel, gasoline, blood – can cause uneasiness as well.

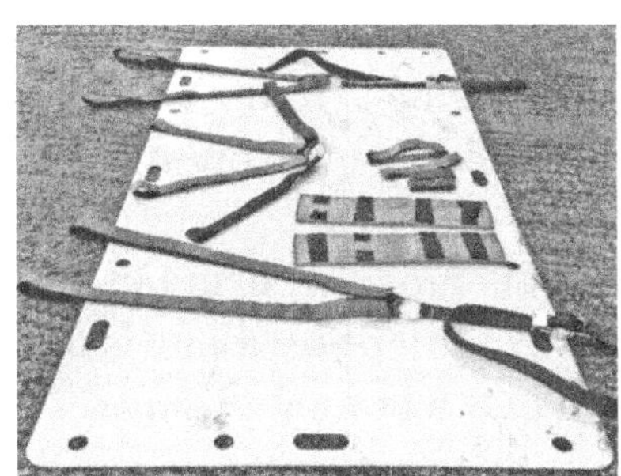
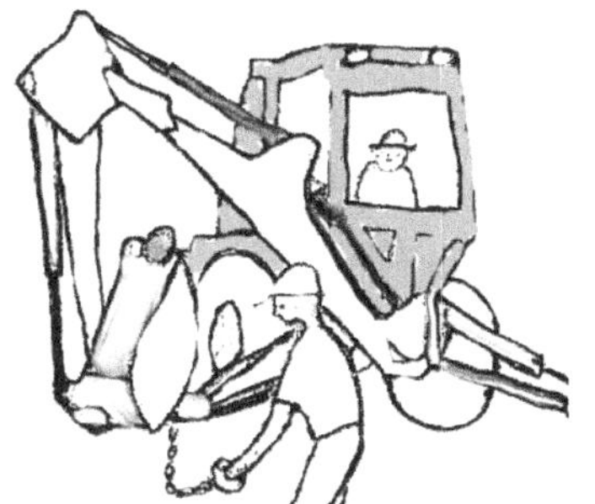
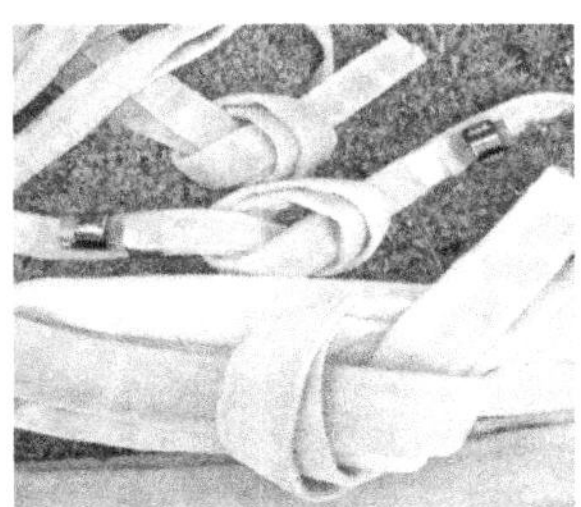

Glide, lift mechanism, and hose
Glide photo courtesy of Resquip; Hose photo courtesy of Vicki Schmidt

From Vicki Schmidt, Maine State Fire Instructor: "Fire attack line, usually 1 ½- 2 ½ inch can be used. Fifty foot sections of "out of service" or retired lines with the couplings removed are easier to andle than hose that is still in service. One hundred foot sections can also be used. If preparing hose lines for LAR, a variety of both sizes is recommended. Tie the hose in a water knot (see SECTION

III: "Knots for Horse Owners"), and secure with duct tape if possible. This is especially handy if the emergency requires the use of in-service hose."

If the owner is on-scene, ask if she has any rescue equipment available. She may have a "go bag" with equipment needed for a rescue.

STEP FIVE:

IF THE HORSE IS WEARING A HALTER, attach a soft ½ inch or ¾ inch rope (Rescue Rope, Kernmantle rope, parachute cord), at least 20-30 feet long. The rope should be long enough to reach from the horse to the handler who is on safe ground. This is called the "lead rope". If the horse isn't wearing a halter, and none is available, use this rope to make an emergency halter. (See APPENDIX: "Rescue Equipment"). Do not use a bridle, even if it is already on the horse. This will result in damage to the horse, and possible damage to the rescuers. There is also a security issue; the straps of a bridle are usually made of leather which stretches, and are easily broken. Horses led by bridles will often throw their heads around to avoid pain from the bit – a piece of metal – in their mouths.

Halter vs. Bridles

STEP SIX:

CREATE A SAFE AREA around the scene for the horse once he is removed from his location. The best equipment to use is portable panels but the likelihood of having them nearby is slim. Five or six foot plastic construction fencing reinforced with rigid PVC pipe and held by rescuers can also be used to contain a horse. Another possibility is to tie ropes and straps/hose around trees from two feet to about five feet in height. Be aware the horse may panic once extricated and end up back from where you just rescued him.

STEP SEVEN:

IF THE HORSE IS STANDING, it may be possible to fill the hole or ravine with material the horse can use to extricate himself. Straw bales topped with heavy planks of wood can be used to "pad" the area, raising the horse to a level from which he can climb out. If the horse is not injured, it may be possible to lead him to a less steep area where he can scramble out on his own.

IF THE HORSE IS NOT STANDING, but appears to be uninjured: Sometimes a horse may have given up and lain down, or has fallen down and stayed there. To a prey animal; down means dead and he may have given up hope of living. Also, if a horse has been down for awhile, the weight of his own body may cause the "down" side to "fall asleep" or become numb. Prolonged recumbency can cause internal injuries.

If you have room, try rolling him over. Work facing his spine, not his feet. With the handler controlling the horse's head with a webbed halter and lead rope, place padding under the head to protect the eye that's on the ground. Position a 25-30 foot rescue strap on the ground on the leg side of the horse.

Two people stand behind his spine and, using your pike poles, slightly lift his lower front and back legs. At the same time, two others slide the rescue strap under the lower legs, positioning it just **above** the knee on the foreleg and just **above** the hock (backward facing "knee") on the hind leg. Once in place, bring the ends of the strap back over the horse's shoulder and rump and, with two people on each end, roll the horse over while the handler supports the head.

Photo courtesy of Hampshire Fire and Rescue Service

If you have the material, manpower and time, try pushing the horse up into an upright position where his body will be mostly off the ground.

Once you roll him up on his sternum, pad the ground around his back with straw bales or other soft material that will keep him upright.

Horse lying on his sternum

Make sure you are all in a safe position when attempting this. The handler can gently pull on the horse's lead rope while several people can push from the backside. Do not stand on the "feet" side or near the horse's head.

If he is uninjured try encouraging him to stand up. Get his attention by yelling, waving your arms or slapping his rear. Some rescuers have squirted water by squirt gun in the horse's face (not his eyes or nose). It may annoy the horse enough to snap him out of his stupor. Once he's up, give him some time to calm down before proceeding. On the other hand, the horse may be injured – including internally – and cannot stand so do not force him.

STEP EIGHT:

ESTABLISH A MEANS OF PULLING THE HORSE UP VERTICALLY If it is impossible to get the horse out of his situation using his own power you will have to lift him out. Use a front loader or large tractor with lifting ability, a heavy duty bipod or tripod that can hang a pulley over the horse, a truck with a lifting arm, a crane or a tow truck. If using a backhoe, it is safer to lift with the scoop side. Even though the bucket on the arm has the advantage of turning once its load is in the air, the typical backhoe may not be rated for the weight of a horse. Use the scoop on the other side which can usually lift at least a ton. If you are using the "arm side", load the scoop with dirt or rocks to keep the backhoe from tipping.

It's ideal to have something that, once the horse is lifted free of the hole, can turn and deposit the horse on safe ground. A bipod is one option since the arc of the bipod will swing the horse out of the hole. Another option is to attach a side rope to the horse (not at his head, tail or legs!), so he can be pulled over to safe ground while he is held up above the hole. Have this in place before proceeding.

Two Point Harness and Figure Eight Harness
Lift photo courtesy of Jim Green, Hampshire Fire and Rescue Service

NOTE: If the horse has fallen into a well, septic tank, or other vertical confined space, a viable lift harness to use is the Wideman configuration. The advantage of the Wideman over other harnesses is that only the top half of the horse is contained but the horse is still supported. Once the harness is on he can then be lifted by conventional methods. The Rescue Strap is centered at the chest, the ends of the strap are crossed over the withers (back) and brought down between the front legs, and then the two ends are passed behind the strap at the chest and out at the top. Both straps are pulled together. **Working in close proximity to a struggling animal is very dangerous. Be sure to work with a buddy and wear a helmet. The horse should be sedated first; apply the Wideman; then anesthetize before lifting to completely relax all the muscles. Some of those spaces are tight! NEVER LIFT BY THE HEAD – by halter or lasso around the neck.**

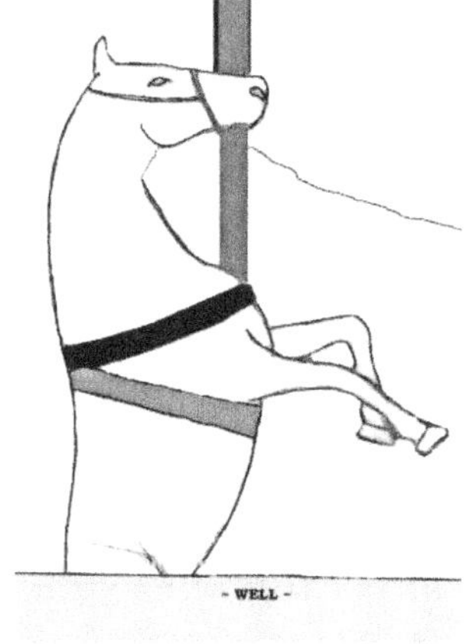

STEP NINE:

APPLY THE VERTICAL LIFT HARNESS If you do not have a commercial harness, make one. (See APPENDIX, "Rescue Equipment" for directions) The harness you will make supports all parts of the horse by distributing the weight evenly, but CAN NOT SUPPORT THE HORSE FOR MORE THAN FIVE MINUTES and is not to be used with a helicopter. Suspending the horse for longer can result in extreme injury. Have the harness on the horse, and everything ready before attempting to lift the horse. Keep the area free of chaos.

For a rescue over rough terrain such as a ravine, where overhead work space is limited, where security is an issue (dropping the horse would be lethal), where injury precludes using a Two Point Harness, or where the harness must be on the horse longer than five minutes, please consider using a more secure lift harness such as the Figure Eight Harness. (See APPENDIX : "Rescue Equipment" for directions) NOTE: Amply pad the legs where the harness comes in contact with sensitive skin (armpits, inside back legs). This is not a comfortable harness and the horse may fidget. Once the Figure Eight harness is in place, start at the beginning and take up all of the slack before tying the final knot.

IF THE HORSE IS INJURED and cannot be righted before extrication, he may need to be dragged from his position. He should be sedated by a veterinarian before this is done. Remember, working around a sedated horse is dangerous; use a buddy!

After the horse is sedated, position the Rescue Glide, Rescue Mat, or its equivalent next to and as far under the horse's back as you can get it. A Rescue Glide looks like a cross between a toboggan and a human backboard and is made of high density, fairly rigid plastic.

When a horse is sedated he will not blink properly. Cover his eyes to prevent retinal damage from wind, rain, debris, or sun. Be aware that the action of covering his eyes may also panic him. Protect his head with padding as well.

If a Rescue Glide, Rescue Mat, or inflatable Rescue Path is not available for dragging the horse, use makeshift material such as heavy tarps or plywood sheets. The material must be sturdy enough to bear the weight of the animal over possibly rough terrain, and be non-shredding so that pull ropes can be attached.

Position the handler at the horse's head to support it, placing padding under the head to protect the ground-side eye and roll or drag the horse onto the "slider". To roll, work facing his spine, not his feet. Position a 25-30 foot rescue strap on the ground on the leg side of the horse.

Two people stand behind his spine and, with pike poles, slightly lift his lower legs. At the same time, two others slide the rescue strap under the lower legs, positioning it just **above** the knee on the foreleg and just **above** the hock ("knee") on the hind leg. Once in place, bring the ends of the strap over the horse's shoulder and rump and, with two people on each end, roll the horse over while the handler supports the head.

Photo courtesy of Hampshire Fire and Rescue Service, UK

Once the sedated horse is positioned properly, tuck his feet into his body and tie them together – front foot to front foot, rear to rear, then front to back feet. Use straps or webbing to secure the horse onto the sliding material. Attach your haul system to the front.

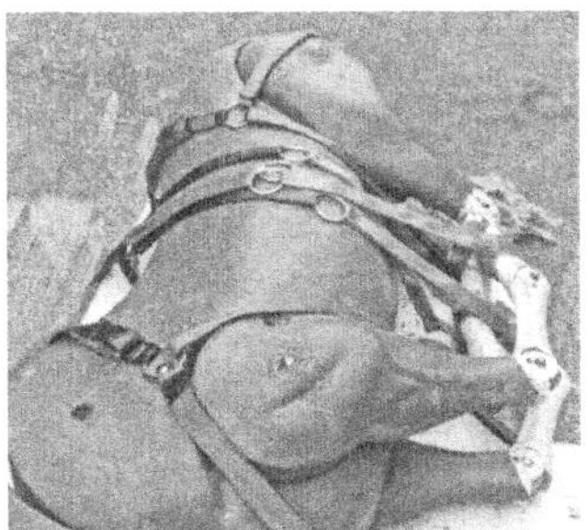

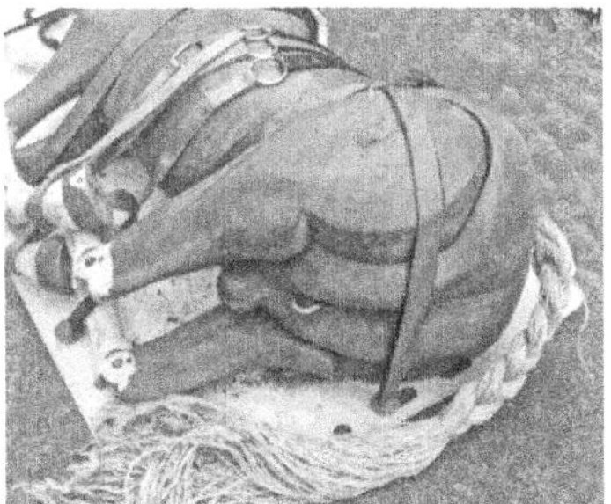

Photos courtesy of Resquip

If you have the proper equipment, and other options are not available, you may be able to lift him by his feet, supporting his back with Rescue Straps or wide webbing. Be sure to secure all four legs and his head. If this isn't possible, do not use this method.

Make sure the animal's head is well padded and supported; tie the horse's head to your sliding equipment so it won't drag on the ground. You can pad his head with sweatshirts or blankets. Be sure to secure the head so it won't come loose while dragging.

Once he is safely out of his predicament, untie him, roll him off the sliding material if it is rigid, and allow him to lie calmly; do not rush the horse to stand up. If you are using a tarp and can't remove the horse from it, tuck it under his shoulder and hip so he won't slide on it when attempting to stand. A horse coming out of sedation can injure himself and those around him trying to stand while still under the influence of the sedatives.

STEP TEN:

LIFT THE HORSE as smoothly and efficiently as possible. Horses usually struggle when they feel themselves being picked up, but once their feet are clear of the ground, they generally relax. Expect a struggle when the feet touch the ground at the end of the rescue. Or the opposite may occur, and the horse will collapse. Stay calm and quiet. Do not keep the horse in the air for more than 5 minutes (Two Point Harness) to ten minutes (Figure 8 Harness).

STEP ELEVEN:

ONCE THE HORSE IS ON THE GROUND, the handler should be prepared for him to panic, and should attempt to stand facing the horse's side so the lead rope can be pulled at right angles to the horse's path of flight. Make sure there is a place to contain the horse so he can begin to settle down. Keep other rescuers clear of the area. When the horse is calm, the vertical lift harness can be removed.

DIAGONAL EXTRICATION

Photos courtesy of Mariposa County Fire District

Sometimes a horse will end up in a ditch or otherwise be in a position where lifting him is not an option. Dragging him a short distance may be the easiest, quickest, safest method of extrication. Examples of instances where a drag would be your best choice are: horse is cast in his stall (too close to the wall and can't get up); standing in a deep ditch (as in the photos above). In other instances, such as sliding off a trail with a short drop, your choice may be to help the horse get himself out.

Always use the safest, simplest, and least intrusive method of removing the horse.

- Self extrication
- Assisted self extrication
- Drag the horse out
- Lift the horse out
- Use power tools to cut the horse out (as in partially fallen through a bridge)

Where to get help:

- Animal Control and Law Enforcement
- Local horse rescues and horse councils
- CART (County Animal Response Team) SART (State Animal Response Team), and DART (Disaster Animal Response Team)
- Large animal veterinarians

Contain horses using the Three Fs: Fencing; Friends (they're herd animals); Food

A large animal in entrapment is a HazMat incident:
A material or substance that poses a danger to life, property, or the environment if improperly stored, shipped, or handled.

STEP ONE:

SCENE SAFETY As with any incident, scene safety is your first concern. Scene safety is different in intensity and scope when you factor in large animals. Check for down power lines, leaking fuel and other hazmat issues, unstable ground and trees, wildlife such as snakes, loose dogs, human interference, and stable access to the area for your vehicles.

SECURE THE AREA Make sure the area around the site is safe for people to approach. Have only one person approach the horse to check the feasibility of rescuing. Approach quietly and slowly so as not to frighten the horse. If the incident involves a ravine, cliff edge or other natural topography, make sure the ground is solid before approaching. If the incident involves "black water", check the air quality and be sure to decontaminate the horse once he's rescued. If there is any doubt about the integrity of the ground, personnel approaching the edge should wear a rescue line and harness, approach from a different direction, or decline to conduct the rescue.

Rescuer safety is your first concern. Keep all non-responders out of the area.

STEP TWO:

CALL THE VETERINARIAN -- a large animal veterinarian who is experienced with horses. There could be more than an hour wait time before a veterinarian can be on-scene. You will need a veterinarian to sedate, and later treat, the horse. The veterinarian can also determine if the horse is so grievously injured he can't be saved, and will have the drugs to humanely euthanize the horse. Having a veterinarian in attendance will decrease the danger to rescuers, as well – a major consideration. If your area mandates it, inform Animal Control of the incident. Please share this book with them.

STEP THREE:

APPOINT A HORSE HANDLER to stay with the horse for the duration of the rescue, talking softly and calmly. This should be the person with the most experience around horses. The horse handler will monitor the horse's physical and emotional condition, interact with the veterinarian, as well as watch for obstacles when moving the horse. Monitor the horse's body language, such as ear direction, to follow the horse's focus. If the owner is calm enough, you might need to use her. Keep in mind that the owner's first concern is the horse, not YOUR safety! Using the owner could be a liability issue; if this person is your horse handler you may want to station her in the warm zone with the Incident Commander and the Veterinarian.

Ask the owner the following:

- Has she been trained in Large Animal Rescue techniques?
- Is she willing to observe her horse from a distance, passing along information about the horse's state of mind?
- Has the horse been sedated (pass along this information to the veterinarian)?
- Are there any known issues with the horse that will further jeopardize your safety? (Hates men; can't tie; likes to kick)

As much as possible, keep other people away from the area. If more than one horse, provide a handler for each.

Find out if the involved horse is insured. There may be existing protocol about handling him. As you await the arrival of the veterinarian, and if it is possible and safe to do so, take and record the horse's vital signs. (See SECTION TWO: "Horse First Aid for Emergency Responders"). Use the wait time to prepare the scene, setting up a safe containment area for the rescued horse.

Once the horse is out, you will be handling him with a lead rope. It is called a lead rope because you will lead the horse with it. **SAFETY NOTE TO HANDLER:** When leading the horse, DO NOT WRAP THE ROPE AROUND YOUR HAND! Standing on the left side of the horse and facing forward, hold the rope under the horse's chin with your right hand, making sure the rest of the rope is not looped, but fold it back and forth and hold the folds in your other hand.

NOTE: **Use the lead rope for leading/directing/containment and to keep the horse from slamming his head on the ground. A horse needs free movement of his head when he is trying to stand up, to walk, and to balance. Don't try to pull him – the steady pressure on his head will not allow him to move in your direction. Allow him to help you in his rescue.**

KEEP THE HORSE CALM This will be your main focus, throughout the rescue effort. Keep the area free of chaos.

- Horses defend themselves by fight or flight
- When they feel trapped, their stress level shoots up. This can be disastrous for their medical condition and makes the danger much greater for rescuers. It is difficult to sedate a stressed horse, it may take up to 8 times the normal dose of sedative, so it is better to keep the horse calm in the first place
- Panicked horses may thrash and strike out with hooves and teeth at anyone who gets close, slam with their heads or hindquarters, or run over or roll over anyone in their way. The calmer the horse, the greater the safety for emergency responders
- Gently stroke – don't pat – the horse (less aggressive). Avoid eye contact or sudden movement. Speak softly in a monotone voice.
- Stallions, and mares with foals, are especially dangerous

AVOID THE DANGER ZONE! – directly in front of horse; up to 8 feet to the rear and to the side of rear legs, depending on the size of the horse.

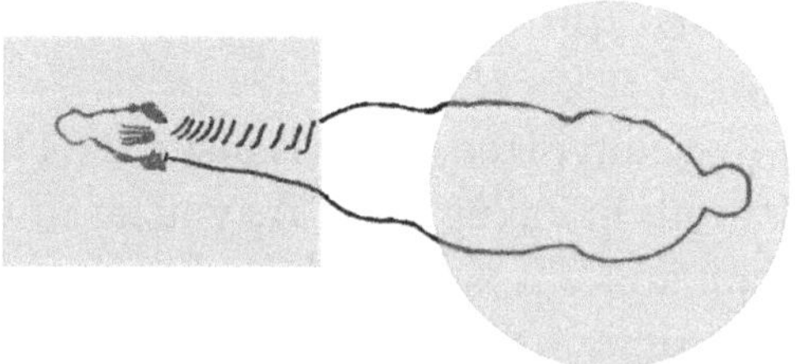

KEEP THE NOISE LEVEL DOWN Extreme light/dark contrasts such as those caused by flood lights used at night are frightening to a horse because he will not be able to see properly. By keeping the area free from chaos and loud noises you will lessen his urgency to flee, thereby lessening the risk to rescuers.

If you use generators to power your tools, move them as far from the rescue site as possible.

STEP FOUR:

GATHER THE EQUIPMENT NEEDED to perform this type of rescue. For a drag or assisting a horse up a steep slope you will need one (assist) or two (drag) Rescue Straps – either commercial, or made from 2 ½ to 4 inch webbing or hose; rope; soft padding material, and a pulley system. If dragging the horse is a possibility (other than a short haul) you will also need to get something to drag him on – either a Rescue Glide, Rescue Mat, inflatable Rescue Path, or something solid and non-shredding like a heavy tarp or plywood. It is a good idea to make a pair of earplugs for the horse from lightweight socks stuffed with cotton and tied at the open end. Muting the sounds associated with the rescue may help calm him. The smell of turnout gear – smoke, diesel, gasoline – can cause uneasiness as well.

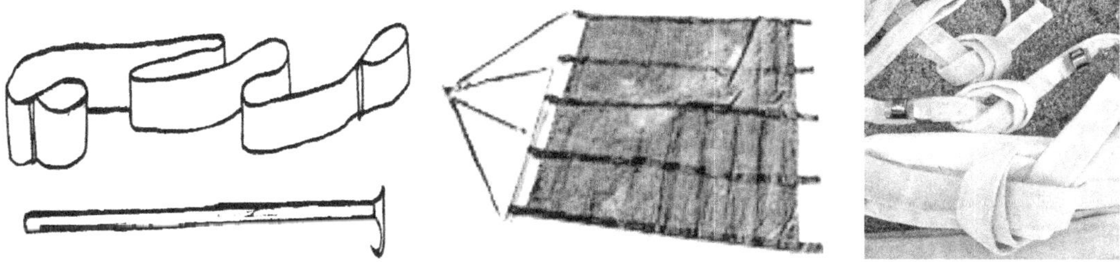

Rescue Strap, pike pole, Rescue Mat and hose
Mat photo courtesy of Timothy Collins; Hose photo courtesy of Vicki Schmidt

> *From Vicki Schmidt, Maine State Fire Instructor:* "Fire attack line, usually 1 ½- 2 ½ inch can be used. Fifty foot sections of "out of service" or retired lines with the couplings removed are easier to handle than hose that is still in service. One hundred foot sections can also be used. If preparing hose lines for LAR, a variety of both sizes is recommended. Tie the hose in a water knot (see SECTION III: "Knots for Horse Owners"), and secure with duct tape if possible. This is especially handy if the emergency requires the use of in-service hose."

If the owner is on-scene, ask if she has any rescue equipment available. She may have a "go bag" with equipment needed for a rescue.

STEP FIVE:

IF THE HORSE IS WEARING A HALTER, have the handler attach a strong, soft ½ inch to ¾ inch rope, at least 20 feet long. This is called the "lead rope" because you will use it to lead the horse. The rope should be long enough to reach from the horse to the handler who is on safe ground. Having two handlers, one to each side of the horse, will be safer as the horse traverses a slope. If the horse doesn't have a halter on, and none are available, use this rope to make an Emergency Halter. (See APPENDIX: "Rescue Equipment" for directions.) Do not use a bridle, even if it is already on the horse. This will result in damage to the horse, and possible damage to the rescuers. There is also a security issue; the straps of a bridle are usually made of leather which stretches, and are easily broken. Horses led by bridles will often throw their heads around to avoid pain from the bit – a piece of metal – in their mouths.

Halter vs. Bridles

It is safer to have a handler with a 20+ foot rope on either side of the horse to control the horse's direction and for handler safety.

STEP SIX:

CREATE A SAFE SPACE TO CONTAIN THE HORSE For an outdoor rescue, once the horse is out of danger you will need a safe place for him to stand, and room for the veterinarian to work if one has been called. The best equipment to use is portable panels but the likelihood of having them nearby is slim. Five or six foot plastic construction fencing reinforced with rigid PVC pipe and held by rescuers can also be used to contain a horse. Another possibility is to tie ropes or straps/hose around trees from two feet to five feet in height. Be aware the horse may panic once extricated and end up back from where you just rescued him.

STEP SEVEN:

FOR A DIAGONAL ASSIST If the horse is standing and has a path which he can navigate with some help, apply the Rescue Strap, either commercial or made from fire hose or webbing, around his chest. To make a Rescue Strap, use a 2½ to 4 inch fire hose, cutting 6 inch slits about one foot from the ends of the hose or tie loops. You can also use 2-4 inch webbing, tying securely knotted loops at the ends of the webbing.

The straps lay on strong bony structures of the horse, avoiding further injury. Least secure is the simple Forward Assist (picture below on left): The Rescue Strap is centered over the horse's withers (back) and two ends are fed between the horse's front legs and are attached to your pulley system. Lark's Foot: The longer end of the strap is pulled through the loop (or slit) at the other end of the strap, so there is a "noose" around the horse's middle. Attach the long end to the pulley system. The Lark's Foot puts significant pressure on the ribs/sternum, but can be considered where extra security is needed. The Wideman Configuration (pictures on right): The Rescue Strap is centered at the chest, the ends of the strap are crossed over the withers (back) and brought down between the front legs, then the two ends are passed behind the strap at the chest and out at the top. Both straps are pulled together. The Wideman is often used for confined space vertical lifts but works very well to assist a horse up a slope. NOTE: it can be safer to have a handler with a lead rope on either side of the horse to control his direction. (They're not as likely to get run over!)

Forward assist straps (Simple and Wideman)
Photo courtesy of Vicki Schmidt

IF THE HORSE IS NOT STANDING, but appears to be uninjured: Sometimes a horse may have given up and lain down, or has fallen down and stayed there. To a prey animal, down means dead. Also, if a horse has been down for awhile, the weight of his own body may cause the "down" side to "fall asleep" or become numb.

If you have room, try rolling him over. Work facing his spine, not his feet. With the handler controlling the horse's head with a webbed halter and long lead rope, place padding under the head to protect the eye that's on the ground. Position a 25-30 foot rescue strap on the ground on the leg side of the horse.

Two people stand behind his spine and, using your pike poles, slightly lift his lower front and back legs. At the same time, two others slide the rescue strap under the lower legs, positioning it just **above** the knee on the foreleg and just **above** the hock (backward facing "knee") on the hind leg. Once in place, maneuver the ends of the strap back over the horse's shoulder and rump and, with two people on each end, roll the horse over while the handler supports the head. This maneuver is helpful when the horse's legs are facing uphill.

Photo courtesy of Hampshire Fire and Rescue Service

If you have the material, time and manpower, you can try pushing him up into an upright position where his body will be mostly off the ground. Once you roll him up on his sternum (breastbone), pad the ground around his back with straw bales or other soft material that will keep him upright.

Horse lying on his sternum

Make sure you are all in a safe position when attempting this. The handler can gently pull on the horse's lead rope while several people push from the backside. Do not stand on the "feet" side or near the horse's head.

If the horse is uninjured try encouraging him to stand up. Get his attention by yelling, waving your arms or slapping his rear. Some rescuers have squirted water by squirt gun in the horse's face (not eyes or nose). It may annoy the horse enough to snap him out of his stupor. On the other hand, the horse may be injured – including internally – and cannot stand so do not force him. Once he's up, give him some time to calm down before proceeding. If the horse is injured or otherwise can't stand, skip ahead to Step #10.

STEP EIGHT:

FASTEN A ROPE TO THE RESCUE STRAP with the other end of the rope attached to a pulley system that is secured to a very stout tree or similar anchor above the "place of safety." The haul team can extend in a line at an angle to the direction of pull, as the pulley will redirect the force to pull the horse up the slope. Never position your haul team in the path of progress of the horse. If your lines run across other ropes, rocks, walls, or other sharp objects, be sure to protect them at the point of contact. Your anchor MUST be solid. One thousand plus pounds of struggling horse can pull people over the edge with him. If the anchor is secure and the horse struggles, he will most likely calm down if you work slowly, quietly and confidently. Make sure you have a breakaway strap between the two that can be quickly cut.

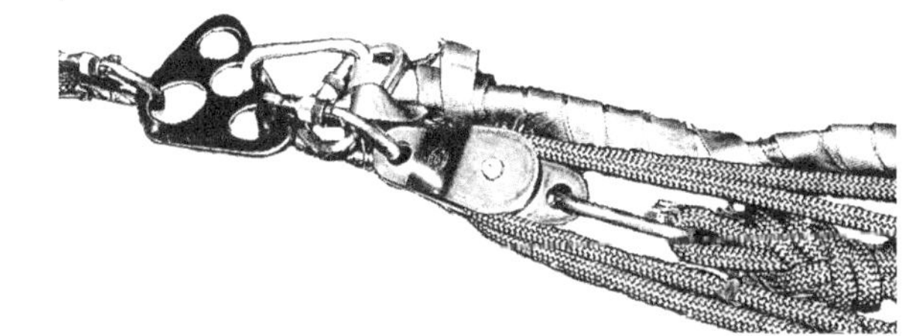

STEP NINE:

ASSIST THE HORSE UP THE SLOPE The horse should be encouraged and gently guided by the lead ropes (one to each side), and assisted by the Rescue Strap, but **the work is done by the horse**. This way, there is no trauma to his neck as he walks up the slope but he is under the control of the lead ropes, which keep him pointed up hill. The strap is intended to prevent the horse from losing ground and he should be able to traverse the slope with slight tension on the strap. Once he reaches the place of safety the tension on the haul line is released and the rescue strap removed.

Be aware that the horse is unpredictable and rig your lines so that they can be cut or released at a moment's notice.

If the horse is unable to traverse the slope, you will need to consider a vertical extrication or a drag. See the VERTICAL LIFT section.

STEP TEN:

IF THE HORSE CANNOT WALK OUT (and you are not considering a vertical extrication) you will need to pull him out. This step calls for sedation by a veterinarian. Remember: working around a sedated horse is dangerous; work with a buddy!

After the horse is sedated, position the Rescue Glide, Rescue Mat, inflatable Rescue Path or its equivalent next to and as far under the horse's back as you can get it. A Rescue Glide looks like a cross between a toboggan and a human backboard and is made of high density, fairly rigid plastic. If one is not available you can use a heavy tarp or sheets of plywood. This will be a lot harder on the horse and on your haulers but will work.

The material must be sturdy enough to bear the weight of the animal over possibly rough terrain, and be non-shredding so that pull ropes can be attached. Position the handler at the horse's head to support it, and roll the horse onto this "slider".

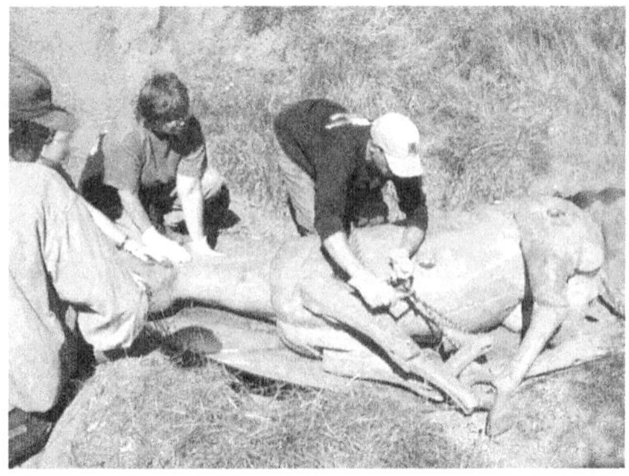

Once the sedated horse is positioned properly, if necessary tuck his feet into his body and tie them together – first back feet together, front feet together, then tie back pairs to front pairs. Some "sliders" can accommodate a recumbent horse with stretched out legs. Use straps or webbing to secure the horse onto the slider. Attach your haul system to the front. The horse's head needs to be supported and protected throughout the pull. You can pad his head with sweatshirts or blankets. Remember that horses are "obligate nose breathers" so don't block his nose. If you are using a Rescue Glide or Rescue Mat, tie the horse's head to the equipment so it won't drag on the ground. Be sure to secure the head so it won't come loose while dragging.

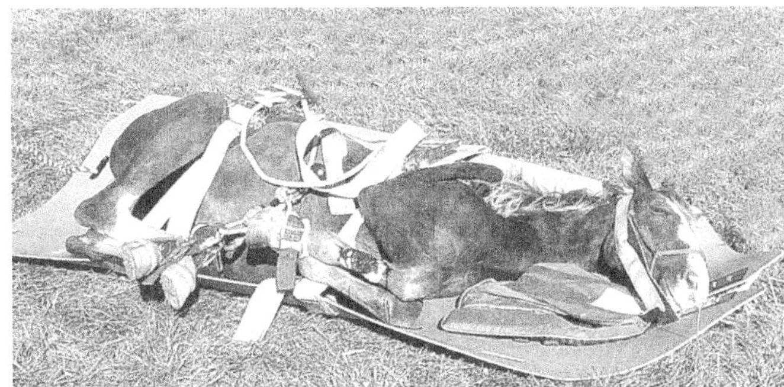

Photo courtesy of Dr. Rebecca Gimenez

If your plan is to transport the horse because of injury, you can then slide him right into a trailer. If he can stand up once he gets to the safe place, roll him off the slider and remove the straps. Allow him to "wake up" from the sedation and encourage him to stand. If you are using a tarp and can't remove the horse from it, tuck it under his shoulder and hip so he won't slide when he tries to stand.

When a horse is sedated he will not blink properly. Cover his eyes to prevent retinal damage from the wind, rain, debris, and sun. Be aware that the action of covering his eyes may also panic him. Protect his head with padding as well.

STEP ELEVEN:

IF THE HORSE IS RECUMBENT AND NEEDS TO BE DRAGGED backward for a few feet, feed two Rescue Straps under him. Have the horse handler kneel on the horse's neck to keep him from trying to rise, and all work will be done from the back of the horse, keeping in mind that the horse's head and neck muscles are very powerful and can inflict a great amount of damage.

Photo courtesy of Dr. Rebecca Gimenez

Like some humans, horses have an "hourglass figure" – they have wide shoulders, and wide hips, with narrow spots next to them. One is just behind the shoulders and the other is just in front of the hips (where a human's would be if he were on his hands and knees). Feed one Rescue Strap under the horse at his withers (behind front leg), then between front legs, under his neck, and back to rescuer. Feed the other strap under the horse at his hips (in front of back legs), then between the back legs, being careful of the placement of the strap especially with male horses, and back to rescuer.

The horse handler is responsible for keeping the horse's head out of debris. If more than four people are needed to pull the horse, double up on each strap. Pull evenly and slowly, watching for an adverse reaction from the horse.

The drag works well for horses who are "cast" in a stall. That means they are down too close to the wall to allow them room to stand up. By pulling the horse a foot or two away from the wall he will be able to stand up. The haul team can stand outside the stall to perform this maneuver, which keeps them safe.

If you feed the straps under the horse at withers and hips, then run the straps back across the horse's body to the haul teams you will have what looks like a "Vertical Lift" configuration. The problem with using this type of configuration rather than the one listed above is that the action of pulling tends to roll the horse over. This leaves the horse with his feet pointing toward the haul teams and effectively stops the drag of the horse.

SPECIALTY RESCUES

QUICKSAND OR MUD

While a common rescue of large animals, this is one of the most difficult and requires careful planning before you start. It is imperative to keep the horse calm as he may sink deeper if he struggles and he is at risk of asphyxiating from the pressure the mud exerts on his ribcage. Pulling the horse out by a halter or a rope around his neck can cause **severe injury**. The method of extrication will depend on how stuck the horse is: use either the "FORWARD ASSIST" (shallow mud) or the "VERTICAL LIFT". Do not use a tractor.

Getting a strap down around his chest will not be easy. When applying any straps you will have to reach down from beside the horse, which is a marginally safe area, due to the drag of the mud or quicksand and the erratic movement of the horse and head tossing. Put some sort of flotation device under his head to keep him afloat. Float pieces of plywood or even backboards on the surface on either side of the horse for your rescuers to stand on.

To get straps under the horse, attach to a light rope to a pole, Nikopolous Needle (a six foot steel pipe, "C" shaped at one end with a loop, and holes along the length for air or water to be dispersed along the horse's legs; developed by Dr. T. Gimenez and Dr. D. Nikopoulos); or even a Strop Guide (six foot steel bar with a "T" handle and a slit at the end, rounded to mimic the curve of a horse's torso; developed by Hampshire Fire Brigade, UK). Attach your strap and pull the strap back toward you.

To ease a horse out of quicksand or mud, inject air or water down his legs, so there is no suction as he is pulled up. You will need to have the horse securely suspended in a lift harness **before** you begin; the injection of air or water will not hold up the horse and can cause him to sink even deeper.

To pull a horse up without relieving the suction can cause his hooves to be pulled right off his legs.

Photo courtesy of Dr. Rebecca Gimenez

SWIMMING POOL RESCUES

A horse who finds himself in a swimming pool presents some unique challenges to rescuers. **DO NOT GET IN THE POOL WITH THE HORSE**.

Draining the pool and building a platform of hay bales topped with sturdy planks is an option to help the horse self-extricate.

If there are steps that the horse would be able to climb, he should be guided by halter and lead rope. You may need to lasso the horse unless you can get him to come to the side on his own.

If the steps are very steep, or the horse refuses to climb, you can try either a "FORWARD ASSIST" or a "VERTICAL LIFT" rescue.

In a "FORWARD ASSIST", the Rescue Strap (or equivalent) is applied over the back, behind the front legs, and the ends then come out between the front legs. These ends are fastened to a rope. On the other end of the rope is a haul team (or rope system) which can pull the horse while the handlers guide him with a lead rope.

Forward assist strap. No strap needed here: horse walked up the stairs
Photo courtesy of Corte Madera Fire Dept.

The strap can be applied by laying it gently over the horse, then using a balloon, soda bottle or semi-inflated ball attached to the end of the strap to pass it under the horse.

The end is pushed under the horse, using a long stick or pike pole (pad the sharp points), and released once it's far enough under so the float will take it up to the surface of the water on the other side. This should be a slow, calm process.

The same can be used to attach a VERTICAL LIFT harness. Once the vertical harness is attached, the horse can be lifted in the same way as if he were in a hole. Read additional information in the APPENDIX, under 'Rescue Equipment": Collins Rescue Equipment.

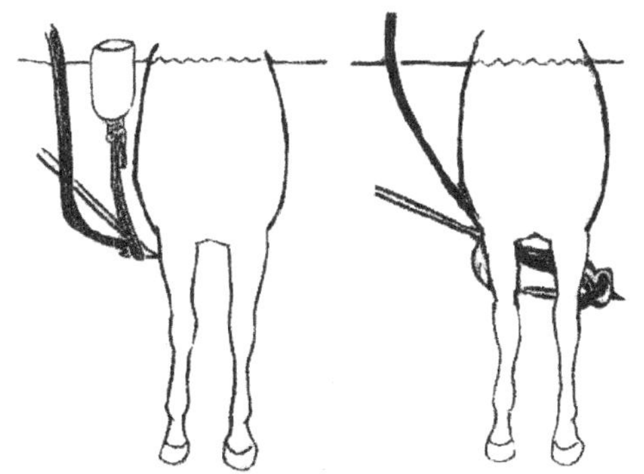

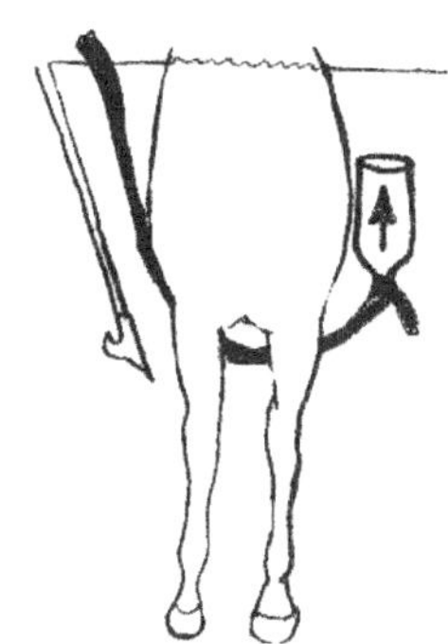

OPEN WATER RESCUES

You may be able to pull the horse in to shore by placing a rope around his neck to contain him and doing a "FORWARD ASSIST". Getting into the water with the horse is extremely dangerous, as his flailing hooves and head can be fatal to you. Hidden obstacles such as waterlogged trees, holes, fencing, or rocks make footing unsafe, and there may be chemicals or sewage in the water.

A horse can swim well, but not for a long time. His body floats but not his head or legs and as you can see from the pictures below his nostrils are very close to the surface of the water (and he's an obligate nose breather). Get a flotation device under his head as soon as possible.

Photos courtesy of Dr. Rebecca Gimenez

If the bank is climbable, use a "FORWARD ASSIST" to help him climb up the bank. Otherwise, use a "VERTICAL LIFT" or a drag to remove the horse from the water. You may be able to get a strap or rope around him by using the Pike Pole/ Balloon Method above.

You need a trained rescue team to attempt this.

SWIFTWATER RESCUES

Swift water rescue is extremely dangerous. It should NOT be attempted if you are untrained. Chances are you will also become a victim and need rescuing. Swift water is relentless, powerful and can kill you. It is predictable ONLY if you know what you're doing. The information given here is for knowledge only.

It's important to understand that, like humans, animals have stress response to diverse situations. Animals caught in moving water may panic. It has also been recognized when possible and safe, in a rescue of a human with an animal, the rescuer should try to remove the animal as well as the human at the same time. When animals are left behind it has been proven that the owners will return to get the animal out. This can lead to doing a double rescue of the same person. It is best, if possible, to have the owner assist in getting the animal stabilized. Be sure to have experienced people on shore. A horse kick can kill a man if delivered to the chest or head.

The lowest tech, least dangerous rescue type should be your first choice. If you can extend a lasso on a pole from the shore to the horse and snare him, this would be your first option. Rescuers stay safe and the animal is brought to the shore. However, this method is obviously very limited in use. Throwing a rope over his neck requires some cowboy skills and repeated tosses may discourage the horse from coming near you. Sometimes just verbal encouragement or offering food will change the course of the animal and he will come close to shore where he can be caught.

If you have the time and space you may be able to string a barrier such as inflated fire hose or "pool noodles" to stop the downstream progress of the horse and encourage him to swim toward shore.

Once you enter the water, either swimming or in a boat, you are stepping up to a much higher level of danger. A boat can be used to herd the horse towards a safe area at the shore. This is a high-risk option to both the horse and the rescuers, but is a common solution. Some hay or grain in a bucket might induce the horse to follow the boat to safety, the horse could be haltered and led by the boat, or the boat can be used to herd the horse toward shore. Swimming around large animals is extremely dangerous and should not be attempted.

If your horse is merely stranded by a flash flood, you may be able to provide him with food and water, leaving him where he is until the water recedes. The danger here, of course, is the horse will take it upon himself to self-rescue, swimming out into the water, possibly away from shore.

Helicopters are usually not an option, despite incredible video footage on "YouTube". They are expensive and many times more dangerous than a boat rescue.

Like other types of water rescue, specialized equipment is required; a helmet and personal flotation device (PFD) are a must. Other useful equipment includes a knife, headband flashlight, whistle, and a wetsuit. A flotation device to keep the horse's nostrils out of the water will help to keep him from drowning.

WATER FACTS

- If you can stay out of the water, do so. If you have to go into the water, dress accordingly. Moving water will draw your body heat out of your body at a rate about 250 times that of air!
- Know the dangers! Swift water consists of more than just water. Debris such as tree branches can float or be lodged under the surface. There may be rocks to trap your feet (or a horse's), or manmade obstacles such as crumbled dams and bridges or abandoned cars. There is also the possibility of hazardous materials such as chemicals, gasoline or sewage, if the water is caused by a flash flood
- The speed of the water is quickest in the middle just below the surface, and the flow of the water will push objects into the middle of the river. A bend in the river is slowest on the inside and the flow of water is pushed to the outside of the bend. The slowest water is at the bottom
- If you think swift water is tamable, consider that the water at the edge of Niagara Falls is knee to thigh deep!
- The most dangerous spot is behind an object such as a rock. The current will push you down against the object and hold you under. Also dangerous is what is called a "strainer" – an object that allows water through but not you, just like a strainer in your kitchen. Think of yourself as human pasta trying to navigate through a colander of rocks or tree branches
- Water passing through obstacles forms a visible "V" that points downstream. This indicates the deepest water and a possible clear path between obstacles. A "V" that points upstream can indicate obstacles under the surface

RESCUE SCENE

- Swift water is very dangerous and the edges are often unstable, steep or narrow. Keep unnecessary people out of the area. Always work with a buddy
- Send a spotter upstream. Upstream is where the water is coming FROM. Downstream is the direction the water is flowing TO. Directions on the water are given from the perspective of looking downstream. The spotter will warn of floating debris and will stop any boat traffic
- Agree on your signals beforehand. Once you're involved with the rescue is not the time to talk about what a signal means!
- If you get knocked off your feet, face downstream with your feet up and use your feet to push off rocks and underwater debris. Look for a clear spot on the bank to land. Don't try to stand up again. If the water is deep, swim to shore
- Set up a team downstream to keep both the victim and rescuers from passing this point and floating away
- Establish a containment area for the horse once he's out of the water
- Decontaminate yourself and the horse once out of the water if "blackwater" is involved
- It is now mandated for type 1 NIMS teams to have at least two animal rescue technicians

Our thanks to
Chris Jonason of Wave Trek Rescue
for her technical help with this section

ICE RESCUES

Photo courtesy of Eric Thompson

There are basic facts to consider when attempting an ice rescue. Eric Thompson of Emergency Equine Response Unit in Kansas has compiled the following list. You will find Eric's information in the Resource Section in the APPENDIX. Eric teaches ice rescue classes.

- Rescuers need to be in cold water Personal Protection Equipment (PPE), helmets, and Personal Flotation Devices (PFD)
- A veterinarian should be on hand to deal with hypothermia, shock, and injuries to the horse
- Put floats on all the gear, such as pieces of "pool noodles", boat hooks, etc. so you don't lose them in the water!
- Use methods that inflict the least amount of stress on the horse and rescuer. Most times cutting a lane is easier than rigging a pull system
- When a horse thrashes, he splashes water over the ice sheet. This can cause weakening of the ice and the ice sheet will start to bend and collapse. Knowing the depth of the water is a bonus.

If the horse is swimming, a guide rope may need to be placed around the horse's neck to keep him to one side of the open hole. This guide rope should NOT be used for pulling the horse out of the ice. Its purpose is simply to keep the horse from swimming away from rescuers during an extrication attempt. Floating straps under the horse to complete a side pull may be an option if cutting an escape path in the ice is not a possibility.

If the horse is standing in the water, you might be able to use a forward assist. Do not try to use a forward assist on a swimming horse. The legs are constantly kicking gear out of the way and entangling the webbing.

- If chainsaws are available use them instead of axes. They need to have large enough engines to easily cut though twelve inches of ice. Not every chainsaw can handle ice

- Cut everything at once and then cut into smaller pieces. Ice chunks can be slid under the existing ice sheet to clear an escape path for the horse
- Chainsaws can be unpredictable in ice and can kick back or drag the operator down toward the chain. Using a safety partner will help provide some balance and escape if needed
- If possible, an Emergency Halter should be placed on the horse prior to commencing rescue. Do not pull on this halter – it is not designed for this type of stress on the horse's head and may cause injury
- It is important to protect the horse to prevent the chainsaw from throwing ice and water onto him
- A visual barrier like a spine board, plywood or a glide will keep the horse from trying to go down the cut lane before the rescuers are ready for him
- Horse should be pulled or led into some sort of confinement away from danger. Portable fencing and corral panels work well
- Side body pulls are preferred with a swimming horse or deep water situation. The edge of the ice needs to be protected so the horse isn't cut by ice as he is pulled out
- A flexible mat may be preferred over hard plastic as ice edge protection when pulling the horse out. Occasionally the hole in the ice isn't large enough to accommodate a glide sheet. A thick tarp can be used instead
- Rescuers need to not only consider edge protection for the horse, but also need to be aware that if the ice sheet breaks they could be dumped into the water near the horse's feet. A horse's foot can get entangled with the rescuer and drag him under water. Even if a PFD is worn it is not enough to keep the rescuer out from under the animal. Boogie boards or foam flotation boards are used to distribute the rescuer's weight in addition to keeping him floating above the midline of the horse.
- Rescuers should have a safety line attached to them via a quick release buckle on their PFD which extends back to a land-based team. The safety line should be clear of snags, and provide a direct line of pull away from the escape path of the horse
- Rescuers should consider the lead line for the horse walking out of the ice extending from shore and attended by a land-based team. If the ice sheet is compromised while the rescuer is exiting with the horse, the rescuer safety team can pull him away from the animal or a self-rescue can be executed without interfering with the lead line of the horse
- Use the Pike Pole/ Balloon Method above to run straps under the horse. The balloon or jug can be rigged with a pull tape or rope attached to two straps for a side pull. A spreader bar will need to be attached to the two pull straps to separate the straps and keep them from pulling together on the belly
- Cover the horse to prevent hypothermia and shock

According to the Oakland County, MI Sheriff's Office website, survey the ice, keeping in mind that ice conditions change day by day, lake by lake and location by location on the same body of water. Some signs of changing ice conditions can be, but are not limited to: moving water near a stream, river, unseen spring or inlet; slushy areas; depressions in the snow; heavy snow; white "milky" or black colored ice; and, "frazzle" ice weakened by the freeze-thaw cycles. Frazzle ice is pocketed with tiny air pockets and often looks like frozen slush. These are all signs of thin ice or unsafe ice.

According to the University of Maine website, "Frazzle" or frazil ice forms as needle-like crystals or thin, flat circles in the water when the surface is "supercooled" to just below freezing. Frazil ice is the beginning of slush, pancake, anchor, and other ice forms. River currents move the ice crystals, causing them to form circles or plates.

Our thanks to
Eric Thompson of EERU
for his technical help with this section

HELICOPTER LIFTS

Another option, should you have the helicopter and a pilot willing to risk it, is to put a harness around the horse and lift him straight up. The helicopter needs to have at least a 1500 lb. sling load minimum, which rules out the smaller, more common helicopters.

There are also huge problems with this. It is usually difficult to find a helicopter pilot who is trained to do this sort of rescue. It is very dangerous, as the horse on the long line is an unstable load, and there are sometimes cliffs – and their attendant wind shears – with which to contend. In other words, any rescue lift may be fatal to helicopter and crew, so most won't try it. If a helicopter is to be used, be sure to use a harness that is designed for this purpose.

Some helicopters rapidly overheat when hovering. The lifting system should be properly attached to the horse before the helicopter arrives and rescuers have to be ready to immediately attach the harness to the helicopter's haul rope. There should be a team standing by wherever the horse is to be landed. This team disconnects the horse/harness from the helicopter, and then takes the harness off the horse.

Since it is a very expensive operation, its use is minimal; mostly rescues out of floodwaters or extreme terrain.

SECTION TWO

This page intentionally blank

HOW TO BE SAFE AROUND A HORSE

In order to be safe around a horse, you need to understand his nature and his limitations. While most horses are used to being handled by people, **A FIREFIGHTER IN FULL TURNOUT GEAR DOES NOT LOOK OR SMELL LIKE A HUMAN.** Talk to the horse to let him know there is a person under all that paraphernalia.

Large animals like horses, cows, sheep and llamas are prey animals. They are eaten by predators. Their dominant response is to flee danger, but horses will fight with legs and teeth. Cows will use their heads first, then hooves. Sheep will butt; llamas will bite, spit and strike with forefeet. Prey animals do not pause to evaluate a threatening situation; they respond to fear by flight or attack with little or no prior warning. Humans, like dogs and cats with eyes at the front of their faces, are predatory animals.

Prey animals instinctively know that if you stop to evaluate a strange situation you will probably be eaten. This is exactly opposite of predatory animals (like humans) who will stop and evaluate a strange situation before proceeding.

Due to their familiarity with humans, horses & llamas can often be led, but they can also be driven by a wily and patient herder. Cows, pigs and sheep need to be driven. There is more information on non-horse species in the chapter entitled "**General Tips for All Other Livestock**".

THE HORSE'S SENSES

SIGHT -- Horses have a complex eye to brain structure. Each eye feeds into one side of the brain with limited crossover. This means that while you may cause no reaction when working on one side of a horse, you may cause a reaction on the other. A horse can see approximately 300 degrees around himself. He has a blind spot of about 3 feet directly in front of his nose and another blind spot directly behind him of approximately 6 feet. Though their sight is mostly monocular they also have a binocular field directly in front of their noses. This aids the horse in seeing at a distance and the food right under his nose. Their depth perception is not good, and you will see a horse raise and lower his head in order to bring his eyes into focus.

Although horses can see almost completely around their bodies, needing only to turn their heads slightly to see fully 360°, objects behind them may appear to be moving faster than they actually are.

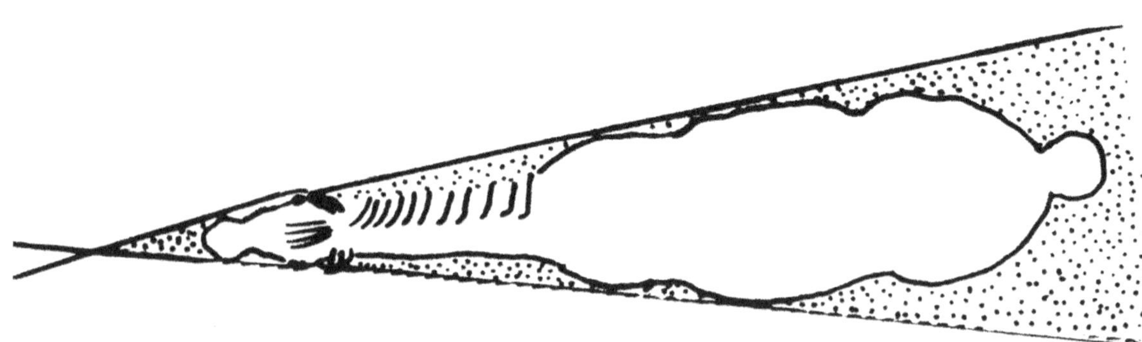

shaded areas are blind spots

Most animals, other than birds and primates, see just two colors (blue and green). The colors animals see best are yellowish green and bluish purple. Therefore, yellow is a high-contrast color for almost all animals. This may be why animals react strongly to yellow turnouts and machinery.

According to Dr.Temple Grandin in "Animals in Translation", horses see the way they do because of the difference in the shapes in their eyes.

Human retinas have a "fovea", a round spot in the back of the eye where they get their best vision.

Domestic animals -- and fast animals who live on the open plains -- have a "visual streak" instead of a fovea. The visual streak is a straight line across the back of the retina. Most experts think the streak helps animals scan the horizon.

Many animals see more intense contrasts of light and dark because their night vision is so much better than ours. Good night vision involves excellent vision for contrasts and relatively poor color vision.

Animals' sharper contrast seems to make dark spots appear to be deeper than lighter spots; the reason cattle guards work.

In "An Anthropologist on Mars", Dr. Oliver Sacks told about an artist who lost his color vision. It became very difficult for him to drive because tree shadows on the road looked like pits his car could fall into. Without color vision, he saw contrasts between light and dark as contrasts in depth.

HEARING – Horses hear extremely well, and their range is much larger than humans. Each ear is independent of the other; they are able to turn in all directions; and they triangulate sound allowing them to pinpoint the source of the sound instantly. The ears aid the eyes in locating objects and all possible danger. The ears are one of the best indicators of a horse's mood. The ears are held forward when the horse is interested, pricked rigidly forward for anxiety or excitement, twisted toward sounds to listen, and laid back tightly against the top of the neck to show displeasure or aggression. If his ears are laid back against his head, use extreme caution working around him until he's calmed down.

SMELL – Horses have a highly developed sense of smell, and they use it to identify objects in their surroundings, remembering other horses, people, places, and things. Allow the horse to smell your hand, your equipment, whatever is causing him to be anxious.

Horses are "obligate nose breathers" which means they cannot breathe out of their mouths. Do NOT cover a horse's nostrils.

SENSITIVITY – The horse is very sensitive to movement and vibration. He can feel the vibrations from your equipment, the traffic driving by on the road, and noise vibrations as they travel up his legs from the ground. This sensitivity warns the horse of unsafe footing or nearby danger. He also has very sensitive areas on his body, including his mouth, feet, ears, and the side of his abdomen just in front of the rear legs.

COMMUNICATION

A horse signals unease by snorting or blowing loudly through his nose. This is ALWAYS a danger signal: flight or fight reaction is imminent. Squealing and ears laid back against the head are signs of aggression – also a danger signal.

Tense heads

HEAD POSITION – A horse will carry his head fairly low when relaxed; carried high is a sign of tension. Chewing or lip licking, on the other hand, denotes relaxation. Watch the jaw and facial muscles for tenseness when anxious; the lips for softness when relaxed.

BODY POSITION – An angry horse may carry his head low, but his body will be tense, his face rigid and his tail will be lashing the air. An injured or sick horse will also carry his head low, but will look hunched up and sunken. A relaxed horse may stand with one back foot resting on the tip of the hoof and will shift from one back foot to the other.

TAIL POSITION – A tail clamped between a horse's legs indicates he feels his life is being threatened. Use extreme caution, as he may feel trapped. A tail that is raised and held up off the body is a sign of alertness and excitement; lashing usually means anger (or biting insects). For simple fly swishing, the tail is relaxed and swishes slowly. Back up your observations by checking other body parts.

HOW TO BEHAVE AROUND A HORSE

MOVE SLOWLY – Reduce the horse's fear by moving slowly. Be sure he knows where you are at all times so he isn't frightened by someone suddenly popping into his field of vision. Any behavior that looks predatory to the horse will usually jump-start the "fight or flight" response.

STAY OUT OF THE "DANGER ZONE!" -- Horses will kick if they are startled by something behind them. If you need to be within kicking range, the safest place to be is close in to his body, so the full force of the kick hasn't developed yet. Otherwise, stay back at least 8 feet.

When standing with a horse, the safest place to be is at his side, by his shoulder. As you're moving around him make sure the horse is aware of your location and intentions. Standing directly in front of a horse (within three feet) will make you almost invisible to him. If you suddenly pop up into his field of vision you could frighten him.

If the horse is lying down, the only safe area is along the spine, away from hooves or head.

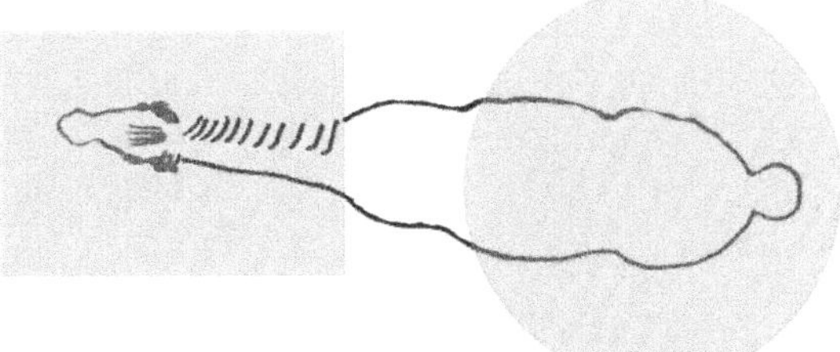

Some horses will rear up and strike with their front feet. Some will run over the top of you. Others will "slam" you with their hindquarters or heads, kick backwards, and all can kick very effectively to the side with their back feet. Be aware that a horse can go from thinking of kicking to connection on the unlucky victim in **ABOUT A THIRD OF A SECOND**!

Traditionally, riding horses known to kick wear a red ribbon attached to their tails. If you notice a red ribbon tied on a horse's tail, be extra cautious.

TALK TO THE HORSE – Most horses can understand some human words, so talk to your horse. Words like "walk" and "whoa", "easy" and "quit" are often known; and the tone of your voice is usually recognized by horses. Use a calm, soothing, confident tone of voice when approaching a horse. Growling may stop him from doing something unacceptable.

Any novel, especially intermittent, high-pitched sounds can cause animals to balk, because such sounds activate the part of the animal's brain that responds to distress calls. Therefore, do not use a "sing-song" voice, but rather a low pitched monotone.

Horses are herd animals. They can be very dangerous in a group as they may exhibit herd behavior with hooves and teeth.

Your position of authority won't last long in a herd, so stay out of the middle. Horses will actually calm down if left in a herd. Horses who were in a trailer together most likely already know each other and will not be fighting for power. The exception is large transports from auctions where many horses are crammed together with no respect for their behavioral instincts.

DON'T ACT LIKE A PREDATOR – Humans are predators. There's no getting away from it. When we're intent on something our body language reflects this fact. Prey animals see this body behavior to mean that they are tonight's dinner. When approaching a horse, or any other prey animal, relax your body. Do not approach in a direct line and, above all, do not stare at him.

HOW TO APPROACH AND CATCH A HORSE

When you first approach a horse, **take the time to evaluate him**. He'll be talking to you with his body – ears, eyes, tail. Evaluate the area as well. If you can approach the horse from a direction that makes use of natural boundaries such as a wall or fence you could cut your walking time considerably! Successfully catching a loose horse requires you to gain some acceptance from the horse. Never "drive" a horse at the person who will be handling the horse. In a herd, when the alpha or dominant horse moves with aggression, the rest of the herd tries to stay out of the way!

If you walk directly and deliberately at a horse, you will look like you are stalking him. Walk slowly, as if you were casually out for a stroll, gradually getting closer to the horse's general area. Watch him out of the corners of your eyes rather than directly watching him. Keep any halter or rope behind your back. Talk to him and praise him. Horses like that. Offering food may work, but usually only if the horse is familiar with you and sees you as his food source.

Try to approach at his left shoulder at an angle from the front, if possible. **Talk calmly to him.** He needs to recognize you as a human rather than a new type of predator. If the horse makes any move away from you, stop approaching and wait. Do not automatically start walking when the horse moves.

Sometimes if you stay motionless the horse will take just a few steps then stop. Remember, any indication of tension and urgency will just convince the horse to move faster. If you time your stride to match the horse's hind feet he may slow down to match your decreasing pace, then may stop when you do.

Once he lets you close, he may face you. This is a sign of trust and his overall appearance will be more relaxed. Take advantage of this trust and back up a step or two. While this seems to be exactly the opposite of your goal of capturing the horse, it removes the pressure of pursuit from the horse. Remember, horses are herd animals and they are vulnerable when they are alone. If the horse moves toward you a few steps, back up a few more steps and wait. He may, then, approach you and nuzzle you, allowing you to stroke his neck. Don't slap him; stroke him gently. Consciously relax your body.

At this point you may be able to calmly put a lead rope around his neck and apply a halter. Don't be afraid to try this technique over and over, or get discouraged if it does not work the first or second time. Depending on the level of panic in the horse, and the chaos of the surroundings, it may take a few tries until the horse feels comfortable with you and you gain his trust.

You may WANT to be aggressive with a horse, either to deflect him from charging you or just to get his attention. To deflect his charge, stand your ground, drawing yourself up as tall as possible. Throw your arms out wide, lean forward and yell. This will scare all but the most determinedly aggressive horses. To get a horse's attention, a loud sound or one aggressive step will startle him into looking at you. Remember to gauge your movement to the situation. Overreacting can undo hours of patient approach.

If you need to approach from the rear, make sure you talk to him so he knows you're coming. Approaching the horse from behind MAY drive him forward, or it could earn you his hooves in your chest!

If there is another person working on the horse, **stand on the same side of the horse as that person**. This allows you to protect her and gives the horse an avenue to move away. If you turn the horse's head to one side, his back end will move to the other side. Pull his head toward you to move his back end away. Push his head away from you to swing his back end toward you.

HOW TO HALTER A HORSE

If the horse is wearing a halter, you can attach a rope to the ring under the chin. There may be "lead ropes" in the trailer or tow vehicle. These are ropes with a metal snap at one end that are used to lead the horse. Consider these points:

- The weakest point in a halter/rope setup is the snap
- A halter, particularly one made of rope, can pull too tightly on the horse, causing damaging pressure on the horse's head, right behind his ears
- If you use a nylon (non-breaking) halter and attach a rope with a "permanent" knot, and then tie this horse to an object, the horse can break his neck or back struggling against the restraint

DO NOT USE A BRIDLE FOR PULLING OR TRYING TO CONTROL THE HORSE. This can cause severe damage to the horse's mouth. A bridle is headgear with a piece of metal called a bit in the horse's mouth. Horses will often fight against the pain or pressure from the bit when being led or pulled by any part of a bridle. There is also a security issue; the straps of a bridle are easily broken.

If the horse is not wearing a halter, and you can't find one, you can make an emergency halter and rope setup. See Rescue Equipment in APPENDIX.

To halter or control a horse, approach from the left side, holding the halter and lead rope in your left hand. Reach your right arm over the horse's neck and slowly move your left arm under his head. Grab the top piece of the halter with your right hand and move it up to the top of his neck behind his ears.

Once this is done, open the halter toward you and slip over the horse's nose. The noseband of the halter should not be resting on protruding bone or on the horse's nostrils or soft area above them. Buckle the halter so that there is some slack at the chin and throat. You should be able to slide two fingers in under the band around his lower face or nose. Too tight and you will restrict his breathing; too loose and he may slip his head out of the halter.

HOW TO LEAD A HORSE

If you need to lead a horse, stay close to his shoulder and walk together toward your goal. A horse will usually resist a person who is facing him and pulling on the rope.

Try to lead and handle the horse from the left side whenever possible. Stand next to his left shoulder to lead him. This is the usual leading position and the horse will probably lead easily from this position. Handling him from the left will usually make him more comfortable.

Hold the rope in your right hand, just under the horse's chin (about 6 inches from the snap at the end of the rope) and keep that arm fairly straight between you and the horse. Fold the remainder of the rope into your left hand. **Do not loop the rope around your hand.**

Ask the horse to "walk", and move forward. Sometimes just using the word "Walk" will cue the horse to move. If he refuses to move it may either be because he can't or what is ahead of him is too scary.

Sometimes you can "unstick" his feet by pulling him to the side before you move straight ahead. You can also try turning him in a very small circle, basically pulling his nose to his tail. You can circle him toward your goal. He may back up better than going forward. You might be able to bribe him with food into moving. Or lead away his trailer partner if he has one. The herd instinct is often stronger than fear of the unknown.

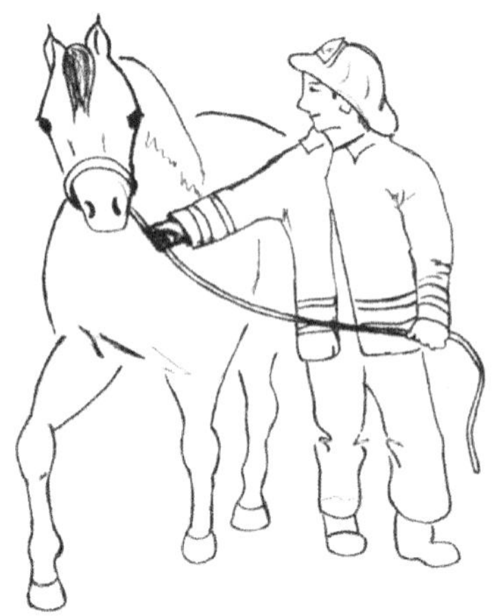

Horses will typically follow their noses. If you can point the horse's nose in the direction you want him to go, he will probably follow with the rest of his body. If you need to turn him, turn him **away** from you rather than pulling him toward you.

Horses will lean into pressure, however, so don't think you can push him to where you want him to go!

If he feels he needs to escape, allow the horse to move around you in a circle, at the end of the rope. This allows him to feel he is running away from danger while you are still in control. It is easier to control a horse by a quick sideways pull than holding onto the rope as he runs away from you. Use the principles of leverage to hold him in a circle. Keep walking toward your goal.

If the horse is circling you, you should be facing him by turning your whole body toward him rather than by passing the rope behind your body.

Is he injured, ill or in pain?

Are you asking him to walk over hoses, toward bright lights, noisy engines?

If it's dark, are you asking him to go from bright lights to darkness, or darkness to bright lights?

NEVER loop the rope around your hand. If a horse takes off, your hand may be trapped *and* you will be dragged

Even a 250 lb. Miniature horse can drag an adult

NEVER hold the rope behind you

NEVER tie the rope to your body or clothing

HOW TO TIE A HORSE

If you can avoid tying a horse in an emergency situation, do so.

Before you consider tying a horse, check the area very carefully.

- Is it far enough away from the incident scene to be calming to the horse?
- Is it free from debris that could injure the horse, such as sharp branches or pieces of metal?
- Is it contained? Out of harsh weather?
- To what will you be tying the horse?

Whatever you attach the horse to needs to be very sturdy and high --- horses are extremely strong. A horse can break a rope if he's frightened enough. Allow the horse to check out the area as you approach it. Use a quick-release knot.

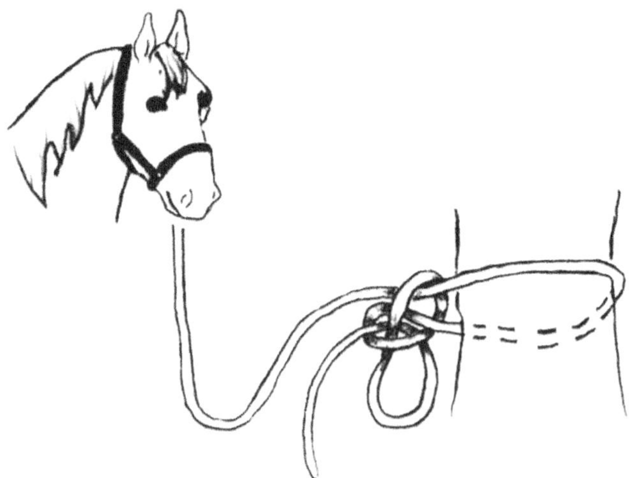

Tie the rope at about his eye-level and with enough slack for the horse to move his head but not long enough for him to step over the rope --- 2 or 3 feet is about right for a horse; less for ponies or miniature horses. Loop the rope around the tree and take another wrap around. Someone should stay near him at all times in case he panics and escapes or hurts himself pulling on the rope. With company, he feels more secure.

<u>YES</u>: Tree, secured trailer
<u>NO</u>: Wire fence, loose fence posts, fence rails, door handles, hitches
<u>NEVER</u>: Tie a horse by the reins of a bridle
Leave a tied horse unattended
Leave a tied horse out in the sun or on concrete for extended periods

HOW TO RESTRAIN A HORSE

Almost any flexible substance can be used to lead a horse if it's strong and long enough to go around the horse's neck – a belt, length of rope or lightweight hose, shirt, pantyhose, or a sack. NOT WIRE. Drape the article around the horse's neck behind his ears. Bring the ends down under his chin.

You can lead a horse this way but only, of course, with the horse's cooperation. If the material is strong enough, cut a hole at the end and feed a rope through it. Tie off the rope and, joining the two ends together, secure with a double half hitch. Or tie a knot in the material and tie a rope to it using a quick release knot. Remember, this will only work as a last resort on a horse that's willing to be led. The best way to restrain a horse that does not have on a halter is to tie your own emergency halter.

Restraining a baby horse, or foal, is an entirely different matter.

Youngsters – like those of any species – are much more fragile. Also, foals probably will not have any experience being restrained. They will often follow their mothers, especially if they are quite young, but if you need to restrain or move a foal, try encircling him with your arms and pull him in towards you. Hold at the front of his shoulder bone rather than at his neck. You can push him along with the arm under his tail or restrain him with the arm around his chest. If you wrap your "forward" arm completely around his chest so that the inside of your elbow rests under his throat, you can raise his head by raising your arm. This gives you more control if the foal begins to struggle. Be sure you do not cut off his air in your enthusiasm. Do not attempt to do this with a larger baby. Chances are he'll leave you in the dust!

HOW TO KEEP A HORSE DOWN

If the horse is recumbent (lying down), and you want to keep him there, put some sort of cushion – a folded blanket perhaps – under his head to protect it -- especially his eyes. Shade his eyes if he'll be in that position for any length of time.

A horse gets up by first swinging his head and neck up, then bringing his front legs forward. At this point his front end will be upright. He then lunges forward and the back end comes up and he's in a standing position. This is the opposite of a cow who raises the back end first, then stands up with his front legs.

Do not stand directly in front of a downed horse's head

Do not stand on the "feet" side of a downed horse

Remember to stay out of the DANGER ZONE at all times (stay behind the lower part of his backbone, above his tail)

NEVER try to force a standing horse down. Both you and the horse WILL be injured, perhaps fatally

When a horse is down you can lean on his neck near his head or even sit on his neck, if necessary. If the horse struggles, you can try placing one hand near or on his head and the other hand on his front leg reaching over from the BACK of the horse, or tip his nose up toward you. You may need as many as four people to restrain a struggling horse.

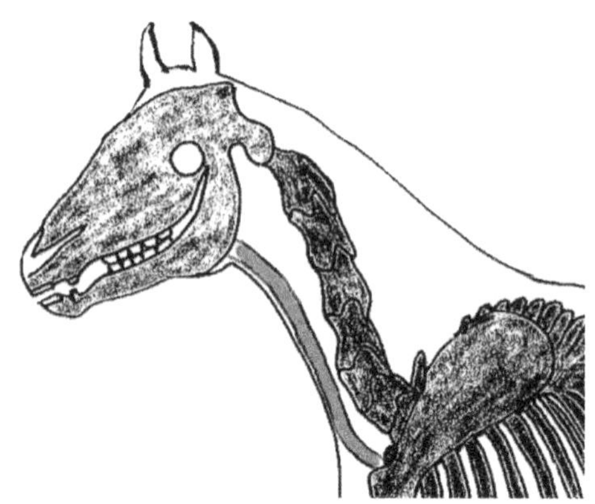

Horses have tremendously strong neck muscles and can launch a grown man in an attempt to get up. Do not sit on his windpipe, which goes along the front and middle of his neck; sit near the top of his neck

Some horses are calmed by covering their eyes. Some horses are panicked. You can try it to see if it helps. Also, do not cover the horse's nostrils. Horses are "obligate nose breathers", which means they cannot breathe through their mouths like humans do.

HOW TO GET A DOWNED HORSE UP

If the horse is not seriously injured, his first instinct is to get up on his feet. This works to your advantage. Make sure the area around the horse is cleared of debris, give him as much room as possible, and then tug on the lead rope. But, beware! The horse may suddenly jump up, ripping the rope from your hands or knocking you down.

You might make "clucking" sounds to encourage him to move. If this is not enough, have someone smack him sharply on his hindquarters (from the "back" side) while you tug.

If this doesn't work it may be that he is psychologically traumatized and has given up trying to save himself. Or, the pressure of being down for a length of time has caused body parts to be numb.

A horse that refuses to get up may have devastating internal injuries. Do not force him.

Do not stand directly in front of a downed horse's head

Do not stand on the "feet" side of a downed horse

Remember the danger zone!

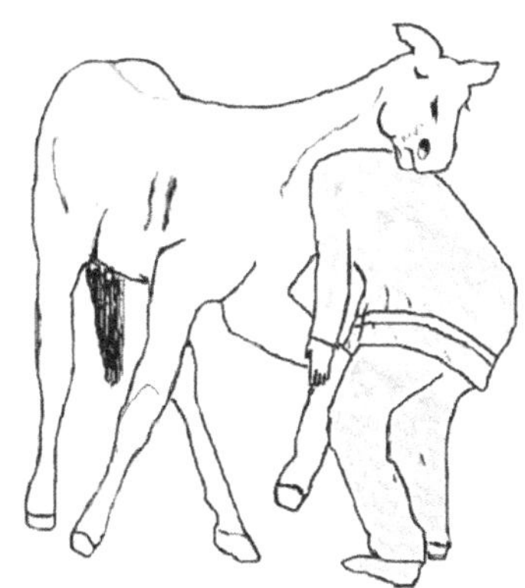

YOU CAN'T GET A HORSE TO STAND BY PULLING ON HIS BODY PARTS!

HORSE FIRST AID FOR EMERGENCY RESPONDERS

CALL THE VETERINARIAN!
DO NOT ATTEMPT TREATMENT UNTIL THE HORSE IS OUT OF DANGER AND IS CALM AND SECURE.
YOUR SAFETY COMES FIRST.

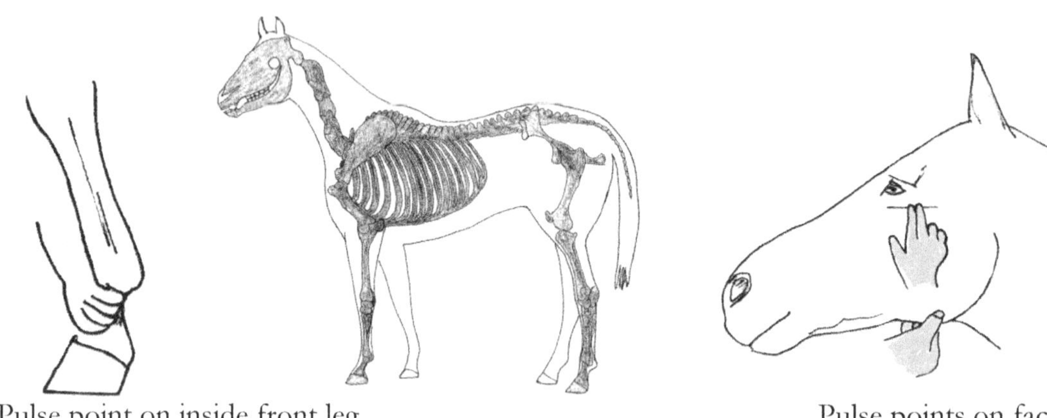

Pulse point on inside front leg

Pulse points on face

VITAL SIGNS

PULSE	30-45 beats per minute; foals are 50-100 beats per minute
RESPIRATION	8-20 breaths per minute
CAPILLARY REFILL	Under 2 seconds

The simplest method for hearing a pulse is to use a stethoscope just behind the horse's left elbow. Remember, the stethoscope will hear TWO beats for each heart contraction while a pulse point has one pulse for each heartbeat.

Capillary refill measures how quickly blood returns to a spot on the horse's gums. Raise his upper lip and firmly press your finger on his gums. This should leave a white spot -- a fingerprint -- that will fill back with pink in a second or two as the blood returns.

Pale, bluish, dry or dark red gums are signs of serious trouble. If the pink doesn't return in less than two seconds, the horse is probably in shock. There is no real first aid for shock. Keep him quiet and put a blanket on him. Shock needs very aggressive treatment from a veterinarian to save the horse.

A few of the serious signs of trouble to report to the veterinarian

- "Sawhorse" or bunched up looking
- Dangling a leg (not to be confused with restng a leg on pointed hoof)
- Sunken eyes, dull eyes, drooping ears
- Lowered head, stiff and grinning face
- Attempts to lie down or roll
- Falling, stumbling, unable to rise, trembling muscles in haunches
- Sweating and kicking at his belly (not like he's kicking away flies)
- Signs of allergic reaction include swelling of lips and muzzle
- Blood in the ear canal
- Bleeding from the nose
- Starvation – you can see his skeleton

WOUNDS

Do not attempt to treat wounds until the horse is out of the danger. Then, use direct pressure to stop bleeding. **Do not remove impaled objects --- leave that for the veterinarian**. Pad really well to protect the object from being jostled. If you remove an impaled object you could easily nick an artery and the animal could bleed to death.

If the horse cannot put pressure on a leg, you may have to splint it. Anyone who has taken a first aid class, whether or not they are responders, knows about splints. Other than size, the only significant difference is that in humans you splint a limb in a "position of use". That means a foot should be in a "walking" position; a hand should not be splinted stretched straight out, but should be slightly curled. Do NOT splint a horse's leg in a "position of use". The toe of the hoof should be down and the heel up. This prevents the horse from putting weight on the leg. Splint only if absolutely necessary.

Use PVC pipe or small lumber pieces; pad between the leg and the splinting material; pad all "hollow" spots where the leg does not meet the splinting material. Use diapers, sheets, blankets, even crumpled newspaper wrapped in a t-shirt – anything soft.

Wrap it all in duct tape. Do NOT attach duct tape to the leg itself. As with human splinting, be sure to splint the joints above and below the break.

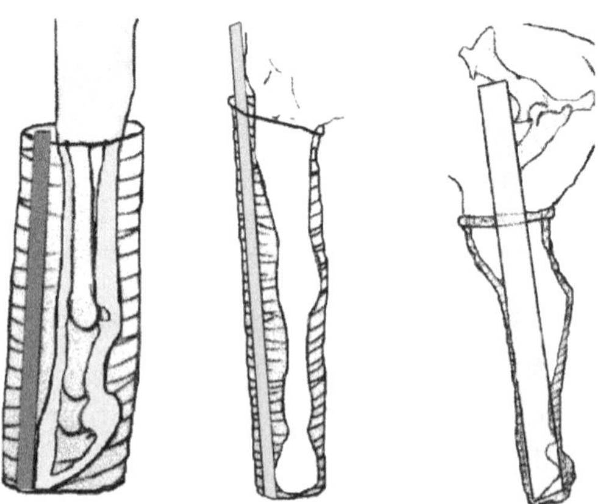

SHOCK

Shock is treatable only by a veterinarian and needs to be dealt with quickly. Cover the horse and keep him quiet. Here are a few of the signs to watch for. As with humans, remove external stimuli.

- Trembling
- Rapid, shallow respiration
- Profuse sweating (cover him for warmth)
- Rapid, weak, "thready" pulse
- Cold ears and extremities
- Weakness and depression
- Pale gums

HEAD INJURIES

As with humans, head injuries can be serious or even fatal. A few of the symptoms are:

- Staggering
- Falling down
- Loss of consciousness
- Obvious trauma to the head
- Great discrepancies when comparing one side of the head to the other
- Blood in the ear canal after trauma can indicate a skull fracture. Additional clinical signs might include severe depression, seizing, and/or holding the head in a tilted manner
- Bleeding from the nose also can be considered a veterinary emergency, especially if the hemorrhage is coming from one nostril and is not associated with exercise.[1]

Keep his head cushioned from the ground if he falls. Keep him calm until the veterinarian gets there. Cover his eyes to protect from the sun and weather.

1 The Horse.com, Article 5618, April 2005

TREATING COLD IN A HORSE (Hypothermia)

Some obvious symptoms are shivering, and cold legs and ears. Cover him with blankets and move him to a warmer, more protected location. As with humans, a super-cold horse will stop shivering, which is a very bad sign.

TREATING HEATSTROKE IN A HORSE (Hyperthermia)

Some obvious symptoms are:

- Profuse sweating including white foamy sweat
- Trembling and muscle tremors
- Rapid breathing and pulse

Spray him with cool water; if there is ice available, place some in a rag or sock and tie to the top of his halter or place an ice pack on the jugular vein to help cool the blood as it circulates. Move him to a shady area or erect a temporary shade cover over him if he can't be moved. Offer fresh water to drink.

DEHYDRATION

Some of the easily recognizable signs of dehydration are:

- Sunken eyes
- Dry, sticky gums
- A pinch of skin that will not fall flat on its own but will stay in a tented position
- Panting
- Dark red gums
- Cold clammy legs

With the vet's approval, give the horse as much water as he can drink. Horses can drink between 15 and 25 gallons of water per day.

EUTHANASIA

Euthanasia may be defined as a gentle death that is free of pain and distress. It is important to select the most humane method in handling and euthanizing the animal, and provide safety for the person performing the procedure and all by-standers.

In addition to using an appropriate method for euthanizing an animal, handling and restraint of the animal in the procedure leading up to its death should be humane and, if possible, it is best to rapidly render the animal unconscious before the moment of death. This would be done chemically by a veterinarian.

Regardless of the method employed, it is absolutely necessary to verify the animal's death by determining the absence of a heartbeat. Some animals may cease to breathe, and then restart if their heart is still beating. The occasional, gasping breath – called "agonal breathing" – is not actual breathing but a reflex. You can confirm death by checking for breathing and heartbeat, and by corneal reflex. To do this, touch the surface of the animal's eye. If there is any response the euthanasia procedure should be repeated.

After the animal dies there may be a period of up to three minutes when the muscles will contract. There may also be some kicking or paddling movements. This is normal.

Experts – and American Veterinary Medicine Association standards – agree that the most effective "field" method for euthanizing all but the smallest animals is by gunshot.

There are two methods: from behind the ear, pointing toward the inside of the opposite eye; and by drawing an "X" on the forehead, between the ears and eyes, shooting above the eyes, as in the diagram. They are equally effective, although the forehead is easier for novices. Do not stand directly in front of a large animal to shoot. They will usually lunge forward as they die. It is preferable to use a large caliber gun, such as a 9mm, .38 caliber, or .44 caliber, although a .22 caliber may work on smaller animals.

Hold the weapon two inches from the head, if possible. Shotguns need to be 12 – 18 inches from the head. Holding against the head could cause a "plugged" effect in the weapon and emotional trauma to the animal; too far away could jeopardize the accuracy of the shot.

Larger, heavier bullets are more likely to be lethal. Rifled shotgun slugs are suggested because of the greater accuracy of shoulder-fired weapons and the reduced potential for the projectile to exit the head. Jennifer Woods, of J Woods Livestock Services in Alberta, Canada, suggests hollow point bullets as the best ammunition.

Make sure the background area is absolutely clear and safe – no people or animals in front of or to the side of your line of fire. Be aware that road surfaces, walls, and metal objects around you can cause ricochet. In polling animal insurance companies, we have found that, while under normal conditions the insurance company has to approve the euthanizing of an insured animal, in an emergency, if the incident commander and/or veterinarian determines the need for immediate euthanizing, the insurance company will honor the policy. This information should help to reassure an already distraught owner.

The insurance companies were quick to point out, however, that the incident must be thoroughly documented with a police report and pictures or an autopsy. Savvy horse owners who carry insurance on their horses will have read their insurance documents and will be prepared.

Stop all traffic in both directions before shooting.

From the rear

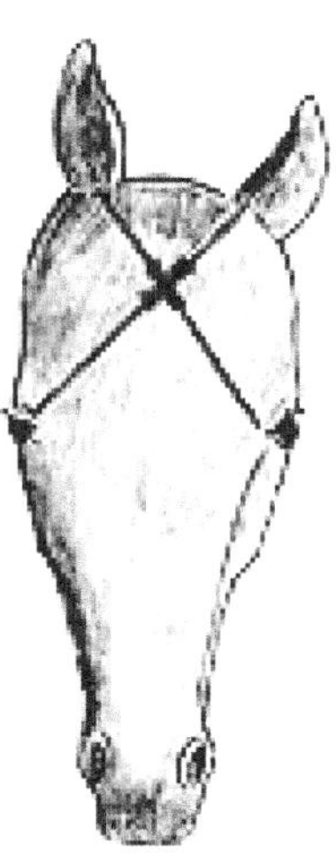

From the front

Quote from Joe Key, firearms expert, on euthanasia:

"Shots behind the ear and slightly forward or through the eye and upward would be the most effective. Rifled shotgun slugs are suggested because of the greater accuracy inherent in shoulder fired weapons and a somewhat reduced potential (compared to other shoulder weapons) for the projectile exiting the head and potentially injuring bystanders. Depending on the round and the skill of the shooter, 9mm/.38 calibers, or higher, would be sufficient. Generally, the larger and heavier projectiles will produce more devastating, thus, more quickly incapacitating, wounds. In all cases, the shooter must ensure that the background area is absolutely clear and safe."

This page intentionally blank

SECTION THREE

This page intentionally blank

CALLING FOR HELP

HOW 9-1-1 WORKS

When you call 911 – or your local emergency number – to report an incident, what happens between that call and the arrival of emergency responders? Areas of the country, and the world, differ in how they handle their emergency calls but most industrialized countries have an emergency number which activates emergency protocol.

In some metropolitan areas 911 calls are initially routed through the Police Department. The Police dispatcher will then transfer the call to the appropriate agency. Some smaller cities have combined their dispatchers, so one central dispatch will be in charge of Police and Fire, and possibly other agencies like Animal Control.

If you call from your cell phone your call may be routed into an out-of-area dispatch center. Sometimes, the dispatch center will take down your information, then will hang up and call the appropriate dispatch center. This leaves you hanging, not knowing if the best dispatch has your information. Often, the local dispatch center would prefer to speak to you to get specific details, such as landmarks or the need for a veterinarian or tow truck. So, if you are calling from your cell phone, **ask the dispatcher to transfer you** to the appropriate dispatch center and not to disconnect you.

In some counties with few people and many miles, your cell phone call might be routed into a dispatch center outside your county. They would need to know your COUNTY, your TOWN, and any landmarks, such as road mileage markers or cross streets. If possible, call from a "landline" (the phone attached to phone lines, such as a pubic pay phone, or the phone "hardwired" into your house).

If the call is for a large animal incident, who responds?

Overturned horse trailer on road:
- Law Enforcement to control traffic
- Animal Control to contain the animals
- Fire Department -- life and property, tools, expertise
- Ambulance (Emergency Medical Services), if humans are injured

Loose horses on the road:
- Animal Control to contain the animals

Horse and rider accident off road:
- EMS for rider; human safety is always first!
- Animal Control to contain the horse
- Fire Department usually responds with ambulance

Horse in a swimming pool:

- Fire Department for their tools and technical expertise
- Animal Control to contain the horse
- Law Enforcement for containment, access to private property

All these instances concern an animal in danger. *Who HASN'T been called?* The VETERINARIAN. In most areas, there is no protocol about animal incidents. So, when you make the call:

- Locate yourself by markers, towns, crossroads -- anything you'd tell an out of town person – and whether vehicles can access the location
- Describe the nature of the incident -- trailer rollover; horse and rider down; horse in water or well
- Indicate the number of people/animals involved and whether they are injured. Indicate if you are traveling with a child
- Specify the need for help for any small animal with you, AND ….
- **Specify the need for an EQUINE vet**
- Express the need for transport if there is no trailer or trailer is broken

The more information you give to dispatch; the better prepared the responders; the quicker the incident is resolved. It is unlikely that the dispatcher will have a list of veterinarians and haulers.

INCIDENT COMMAND SYSTEM

The Incident Command System (ICS) – or your equivalent – is a standardized on-scene management system designed to provide efficient and seamless structure to an incident. ICS combines personnel, facilities, equipment, procedures, and communications within a common organizational structure, and is designed to aid in the management of resources during an incident. It is flexible; it can be applied to small as well as large and complex incidents. When emergency responders arrive on the scene of an incident, they will set up this structure for their response. The ICS structure covers:

- **Common Terminology** – everyone speaks the same language; identifies equipment by the same name; identifies positions by the same term
- **Chain of Command** – There is an orderly line of authority within the ranks of the system with lower levels subordinate to, and connected to, higher levels
- **Unity of Command** – Unity of Command means that every individual has a designated supervisor, and each individual reports to only one designated person
- **Unified Command** – Allows all agencies who have responsibility for the incident to jointly develop a common set of objectives and strategies for the incident. For instance: Your trailer rolls over on a highway at a county line. Responders might include Highway Patrol, local Sheriffs, Animal Control, and Fire Departments from the two counties

INCIDENT COMMANDER

The Incident Commander (IC) is the individual responsible for all activities at an incident, including ordering additional resources and developing a plan of action. "The IC has overall authority and responsibility for conducting incident operations and is responsible for the management of all incident operations at the incident site." *(FEMA: ICS-100b: Introduction to ICS – a free class)*

In a small uncomplicated incident, all positions might be performed by a single person. The first person on the scene of an incident is considered the IC until a person of higher authority or more experience arrives. For instance: Your horse falls into a sinkhole. Once you determine you cannot remove him, you call 9-1-1 for help. YOU are the IC until the responding agencies arrive. At that time, the first responder to arrive will assume command of the incident.

Three important functions are the responsibility of the Incident Commander. They can be performed by the IC, or they can be delegated to others who will report directly to him.

- **Safety** – The Safety Officer's function is to assess hazardous and unsafe situations, and develop measures for assuring personnel safety. This is the person who oversees the safety of the operation, including the rescue personnel, the equipment, and the methods used
- **Information Officer** – The "news person". The central point for dissemination of information to the news media, agencies, and organizations
- **Liaison Officer** – The "go-between". The point of contact at the incident for personnel from other, assisting, agencies

Transfer of Command – When the IC steps down, who will take his place? The preferred method of transferring command is face to face. The preferred person is one of equal or greater experience and expertise

Span of Control – How much responsibility does any one person have? How many resources, including how many people, are under the command of any one person? Typically, any one person will be responsible for three to seven others

Plan of Action – A plan can be oral or written, and contains objectives that reflect the overall strategy for conducting the incident. It can identify resources and the assignment of staff. As additional resources are committed and the incident becomes more complex, separate sections may be established

GENERAL STAFF

A group of personnel organized according to the function they perform. The head of each group reports to the IC. Groups might include:

- **Logistics** – These are the "getters". Responsible for providing services, material and facilities to meet all incident or event needs (personnel and equipment)

- **Operations** – The "action guys" or "doers". Responsible for the direction and coordination of all incident tactical operations. All incident needs (such as personnel, equipment, etc.) are requested from Logistics
- **Finance/Administration** – The "payers". Responsible for monitoring incident-related costs; administering any necessary procurement contracts
- **Planning** – The "thinkers". Responsible for the collection and evaluation of information about the incident; preparing status reports; maintaining status of resources; developing an Incident Action Plan (IAP); and preparing incident-related documentation

Operations for a Large Animal Rescue, or "LAR", could include:

- **Animal Handler** – The handler is the person who is in charge of the animal during the rescue. This person would talk with the owner to get relevant information about the animal; would talk with the veterinarian and report back to the Incident Commander about the medical needs of the animal; and would monitor and interact with the animal during the rescue. Each animal would have a handler
- **Extrication Team** – This team is responsible for the set up and implementation of the removal of an animal from entrapment. This includes hauling, lifting, and scene lighting
- **Containment Team** – This team captures and contains loose animals and is responsible for control of animals after their rescue. They will probably be responsible for providing tarps and blankets to protect the horse from inclement weather

Responding agencies could include:

- **Animal Control** – Is usually the legal authority on-scene. Animal Control is responsible for animals if there is not an owner present; also transports, houses and cares for animals after a rescue

- **Law Enforcement** – Is the "law and order" authority on-scene. This could include police, sheriffs, Highway Patrol or park rangers. Law is responsible for traffic control; crowd control; access to the scene of the incident; road closure if needed; and euthanasia by gunshot, if needed

- **Fire Department** – Is the **main resource** for equipment, manpower, and technical skills.

Other authorities:

- **Veterinarian** – In most cases, the veterinarian will be a large animal or equine specialist, and will be the medical authority on scene. She will usually make the decision on whether an animal can be moved and whether he needs to be euthanized
- **Owner** – Is the ultimate authority on the animal. The owner has the final say in what happens to the animal

Our thanks to retired
Contra Costa, CA Fire Protection District Training Captain, Ana Bertero,
for her technical help with this section.

SAFE TRAILERING

While a lot of this book is geared toward emergency responders, their safety and successful recovery of your horse, this chapter is aimed at you, the horse owner. You can make a big difference in whether or not your horse arrives safely at his destination.

All directions in this section are given for North American driving (on the right side of the road). If you drive on the left, substitute accordingly.

BUYING A TRAILER

The most important consideration when buying a trailer is your horse's safety. This means taking into account his comfort, as an uncomfortable horse can become agitated, causing the trailer to swerve, thereby causing a wreck. Here's my own personal experience with this.

> I bought an old steel 2-horse straight load trailer to use with my Jeep pickup truck. When I bought the trailer I asked the salesman if the truck was appropriate to pull it. "Sure," he said. "No problem. That truck is rated for the combined weight of the trailer and the horses."
>
> I hauled my horses a couple of times with no trouble. Then, one day a friend and I decided to trailer our horses to the beach. On the way home, as we started into a slight bend in the road, one of the horses scrambled in the trailer.
>
> We heard the racket, and I had just enough time to check my rearview mirror for the source of the noise before the trailer swerved violently pulling my truck into the oncoming lane. Fortunately there were no other cars in the vicinity, and I quickly got the rig under control. It was a case of "the tail wagging the dog."
>
> The moral of the story is that just because a truck has the capacity to haul a certain weight does not mean it is heavy enough to keep a trailer on the road when the unexpected happens.

TRAILERS and HORSE HEALTH

When you think of it, the fact that our horses trust us enough to go into a "cave with unstable footing" is somewhat of a miracle. The horse is a prey animal designed to live in open areas where he can see predators approaching from a distance. He has a very finely tuned "flight or fight response", and will run at the least threat. Also, to be able to flee he wants solid ground under his feet. Please be grateful for your horse's trust and don't take advantage of it.

With this in mind, look for a solid, sturdy trailer with a light colored interior and top (or paint the one you have at home) and plenty of windows and vents.

- Large animals pass a lot of air through their lungs, especially when stressed, so good ventilation is a must
- Inevitably there will be a buildup of manure and urine which produces methane gas. Keeping it enclosed in the trailer is not good for your horse's lungs

Horses' large bodies produce heat which can become overwhelming in hot weather, especially if there is more than one horse in the trailer, and hot summertime roads will cause metal floors to heat. Wooden floors are cooler. You can also line the floors with heavy, slip proof mats designed for trailers. Please be sure the mats fit the trailer floor.

Roof vents and open windows will help eliminate fumes and heat. But, if you drive with open windows or are using a stock trailer without screening, be sure to put a fly mask on your horse. Bugs, road dust and small rocks hitting your horse's eyes at 50 mph are sure to generate vet bills! (Same for your pooch!) And, hay wisps will float through the air with the greatest of ease when the windows are open!

Never allow your horse to put his head out the window when the trailer is in motion. Horses have had their eyes gouged, necks broken, and have even been decapitated!

Speaking of things hitting your horse, there should be nothing sticking out into the "box" of the trailer that will harm your horse. Any latches should fold back against the wall with no sharp edges. All butt bars or chains and chest bars should be padded and "quick release", and should be in good condition. All partitions should be smooth and easily removed. Any light fixtures should be well protected from horses' heads.

FITTING THE TRAILER TO YOUR HORSE

Photo courtesy of Dr. Rebecca Gimenez

Consider the size of your horse when looking at trailers. The average 15 hand horse needs about nine to ten feet of length. Warmbloods, drafts, tall, and long horses will need more length, plus extra width and height.

When measuring your horse's *width*, take into account that he will want to spread his legs to steady himself. He will step forward and back, and shift from side to side. Your horse should not have to sque-e-e-e-ze into the stall of the trailer. Measure him across his widest part, and then add a few inches to each side.

When measuring your horse's *height*, do so from an "alert" position. If your horse is startled once inside the trailer and he tosses his head in the air, can he do so without "beaning" himself? Get a poll cap and use it every trip!

According to Cherry Hill's book, "Equipping Your Horse Farm", the inside height of trailers ranges from 72 to 90 inches, with most near 84 inches today.

The height inside an old style horse trailer is 72 to 78 inches. Newer standard sized trailers range in inside height from 80 to 86 inches but it is not uncommon to find Warmblood and draft horse trailers as tall as 96 inches inside. Be aware that the height at the entry might be 4" lower than that at the ceiling and the roof's support ribs are a height somewhere between the two.

Whether you are shopping for a new or used trailer, always take a tape measure with you. Know the outside height of your trailer including open vents and roof rack so that when you pull under a marked overhang, such as at a motel, you don't rip the roof off your trailer. Measure to see if it will fit in your storage building.

A 16-2 hand horse stands 66 inches at the withers. Loaded into a 78 inch trailer, he would only be able to raise his head 12 inches above his withers when loading or unloading and would have to carry it that low through the entire trip. A 16 hand horse would be much more comfortable in a trailer with an 84 to 90 inch ceiling.

A 15 to 16 hand horse can fit into trailer with ceilings lower than 84 inches providing he is levelheaded about loading and unloading, but an 84 inch or taller trailer would be best. If you are planning to haul a very large horse, you may need to look into a custom trailer or van.

Length

In a straight load, length usually refers to the stall length, the distance from the butt bar to the chest bar, and so corresponds to the horse's length from chest to tail. Straight load trailer stall lengths range from 66 to 88 inches with the average somewhere around 70. The average 14-2 to 15-2 hand horse measures about 66 to 68 inches from chest to tail so a stall 72 to 78 inches long (from manger to butt bar) works well for many horses.

In a slant load trailer, stall length corresponds to the length of the horse from nose to tail. The stall length in slant load trailers ranges from 90 to 96 inches. The length of a slant load stall should be measured down the middle of the stall from one wall to the other to correspond to the horse's nose to tail measurement.

A 15 to 16 hand horse is about 92 to 94 inches from nose to tail when the horse's head and neck are in a natural, relaxed position (not compressed or stretching). Horses over 16 hands are at least 96 inches long from nose to tail. Since few slant load stalls are longer than 96 inches, they generally are not suitable for horses larger than 16 hands.

For stock trailers, allow eight feet of interior length for every two horses, so a 16 foot trailer can hold four horses, a 24 foot trailer can hold six horses and so on.

Width

The width of a single trailer stall can range from 26 to 38 inches with most measuring toward the low end.

The stalls in slant load trailers are generally narrower than those in straight loads. Warmblood trailers stalls can be as wide as 48". The average 15 to 16 hand saddle horse does well in a standard width stall as long as the stall divider does not go all the way to the floor. Some horses travel better in a narrower stall, which supports them better, than a wide stall where they can move from side to side excessively.[1]

When measuring your horse's length, take into account his head and neck. He will need to move them for balance. The space for your horse's body should accommodate him comfortably. This is definitely a consideration with slant-load trailers which are built for smaller sized, shorter-backed horses. The practical space in a slant-load stall is often less than 9 feet.

The typical straight load trailer has a shelf [manger] at the front of each stall for hay. If the trailer is too short for your horse, he will have to stand with his head over the hay. This will have two negative consequences. He will constantly be breathing in dust and mold from the hay; and, he won't be able to lower his head to blow, cough

---

1 Equipping Your Horse Farm, Cherry Hill, Storey Books, 2006. http://www.horsekeeping.com

or sneeze. There are also straight load trailers with a bar in front of each stall instead of a manger. How much more healthy for your horse – he can lower his head and blow instead of riding with his nose in his hay during the whole trip!

And then there are wheel wells. There are some trailers that are constructed with the tires under the body of the trailer.

The "hump" inside can cause problems when loading and unloading your horse, and the corresponding stalls in slant-loads will be shorter.

Do yourself a favor and take your tape measure and horse's measurements with you when you shop for trailers.

TYPES OF TRAILERS

When it comes to types of trailers, there is one for every need: Single straight load, tandem straight load, the most common -- side-by-side straight load, center entrance, stock trailers, slant-loads, with or without dressing rooms, living quarters, tack rooms. There are "bumper" hitches, goosenecks and 5th wheels, and vans. Your budget will more than likely determine your trailer purchase!

If money is no object and you plan on hauling multiple horses over long distances, a van may be the answer. It has the most comfortable ride for your horses. Vans have better suspension systems to absorb vibrations underfoot, usually have insulated walls to keep out the din of traffic and heat, have more headroom, and the horses face each other across a center aisle.

Here are some considerations from a safety point of view.

- A "bumper pull" trailer DOES NOT HITCH TO THE BUMPER OF A VEHICLE. This common, but incorrect, term for a straight pull or "tag-along" has been taken literally with dire consequences
- Of the various types of trailers, the straight pull type is the most likely to sway
- Of the various types of hitches, the 5th wheel or gooseneck setup is most stable for hauling 4 or more horses
- The newer suspension systems give a more comfortable ride. This translates into a calmer horse who does not move around as erratically
- Horses riding loose in a slant-load or stock trailer without the stall dividers in place may cause the trailer to sway
- The safest dividers are removable by a quick release mechanism. If the divider is a narrow panel, a horse can fall underneath it. It's very difficult to get a fallen horse out of such a trailer if the divider cannot be easily removed

MYTH vs. REALITY OF SLANT-LOAD

Slant-load trailers are very popular and probably eclipse straight loads in new trailer sales. One of the reasons is that horses are supposed to ride more comfortably in a slant position. Physics will tell you that it isn't necessarily so. If the trailer is accelerating and decelerating in a straight line, and the horse's body is at an angle, the pull on the horse is diagonal. This means that the horse will lean into acceleration with his left rear leg and will lean into deceleration with his right front leg.

Over time and miles, this may cause your horse to be stiff and out of balance, especially if he already has trouble balancing in a too-small stall. Horses riding in a straight load trailer seem to absorb motion more equally.

Horses seem happier to load into a slant-load vs. a straight load but that may have nothing to do with the angle of their bodies. It could have more to do with the openness of the trailer, having a window to look out, and the slant-load may be newer than the straight load and therefore have a better suspension system.

RAMPS vs. STEP UP TRAILERS

The method of entry into a trailer is a matter of preference. There are pros and cons for each, and some safety issues to consider as well. If your choice is a trailer with a ramp, be sure the ramp is not too heavy for you to lift. Here are some points to consider.

Ramps

- Are easier for old, injured and small horses to negotiate
- Need to sit on level ground to prevent wobbling
- Can sound hollow which may frighten some horses
- Horses can fall off the sides, causing serious injury
- Single ramps are more dangerous than double ramps
- If you pick up the ramp from the side, there is a risk of being kicked in the head
- In the event of a rollover, unaware rescuers can be seriously injured by spring loaded ramps

Step-Ups

- Can be difficult for old, injured and small horses to negotiate
- Horses backing out of step-ups have been known to slide under the back of the trailer, causing grievous injury
- Can be intimidating for horses not used to loading into a trailer
- Horses can hit their shins on the edge, causing injury or reluctance to load

- The back door(s) may be used to form a "chute" to direct the horse into the trailer. This is useful for untrained and unwilling horses

From a rescuer's point of view, a ramp can be both more helpful and more dangerous than a step-up. If the rescuers are inexperienced or not trained to remove a horse from an overturned trailer, they may open the ramp like a car door and let go of it. A spring-loaded ramp will slam back into them, causing injury. If, on the other hand, the trailer is upright and the horse inside is injured and recumbent, the ramp may serve very well to slide the horse out of the trailer. Knowing what I do about rescue and trailers, I would choose a step-up, but that is strictly personal preference.

THE SAFE TRAILER

If you are shopping for a trailer and have a limited budget, it's better to get a good used one than a mediocre new one. What determines a good trailer?

From the point of view of rescuers in the event of a trailer wreck, there are many points to consider.

In some older trailers the wooden floors are not attached. If the trailer flips over the floorboards can come loose, injuring the horse inside and possibly dropping his legs outside the trailer to be vulnerable in the rollover.

If the trailer has a rear dressing or tack room, and the trailer rolls onto its left side, either ramps will have to be constructed to slide the horse over the tack room or the tack room will need to be dismantled. If the trailer flips onto its right side, the tack room will need to be removed or horses will need to be hauled out UNDER the tack room.

In a slant-load or stock trailer, if the horses are tied, and if the trailer flips onto its left side, the horses will be lying on their heads. If the trailer flips onto its right side, the horses will be hanging from their ties until they break. If the ropes and halter are stout enough they can have a bungee effect, possibly breaking the horses' necks.

In a straight load trailer, when the trailer flips on its side, if the divider is sturdy, well constructed and stable, there is a good chance the horses will not fall on top of each other but will be kept apart by the divider.

If wooden floorboards are not maintained there is the possibility they can fall through during travel. There are horror stories about horses who have fallen through trailer floors while traveling on the freeway. You can imagine the rest of the story. Check for rot in the boards by probing them with a knife. Spongy boards are dangerous.

If you use mats inside the trailer, be sure they fit snugly. A scrambling horse can move mats into a tangle if they are too small for the trailer.

The trailer should be designed so that each horse traveling within can be reached separately and, ideally, each horse can be removed without having to move other horses to do so.

Picture your trailer on its side with your horses inside. What would YOU like to see to help you get the horses out safely? A window near their heads – protected so legs or heads won't fit through, but where you can comfort your horses while rescuers work to free them? A window where a large animal veterinarian can reach through and sedate the horses? Quick releases on moveable parts? A trailer that is structurally strong enough to withstand rollovers? Use your own questions to help you select your next trailer.

It's a good idea to develop a checklist to be used before every trip and after every stop.

- Are the lights hooked up and working? Check turn signals and brake lights
- Are the brakes hooked up?
- Is the trailer level?
- Are the safety chains crossed and properly hooked?
- Is the battery for the breakaway braking system charged? Are the tires in good condition and properly inflated?
- Check tires for rot if the trailer has been standing idle for awhile
- Are all latches properly latched?
- Is the trailer hitch securely in place and locked?
- Do you have the proper sized ball for the hitch and is it securely seated and locked?

Roxanne Stephens, who drove the "recovery" trailer, tells this next story that happened on Route 175, a busy single lane highway near Chincoteague, VA. It was a cold and foggy February night.

> As a horse trailer was being pulled on this road the ramp came down and a horse fell out onto the road. The driver didn't realize the horse was gone until she reached her destination, 30 miles away.
>
> The speed limit on that road is 55 MPH, but there are a lot of homes along the Route and there are cars turning into their driveways so the speed of the traffic was decreased.
>
> Curious neighbors lured the horse off the road into their small front yards where the grass was green and inviting.
>
> According to the driver, the horse was untied in a 2-horse straight load and she felt he was safe because he was so confined. She hadn't planned on the door latches breaking and the ramp falling down. The horse suffered a lot of scratches, cuts and road burns but nothing serious enough to need stitches. He was subsequently loaded into another trailer and taken home.

Personal experiences have made me very cautious about checking.

> A close friend was traveling with his non-horse trailer. He stopped at a restaurant for lunch and when he came back out just jumped into his truck and started to drive away. The trailer immediately came unhitched because someone had stolen the pin holding his hitch in place. There is now a lock. I took our flat-bed trailer to pick up hay. I drove up the coast road --- winding along cliffs --- and into the hayfield to a barn. We loaded two and a half tons of hay onto the trailer; I drove out of the bumpy field and back along the coast road the 30 or so miles to my house. I backed in to the barn area to unload the hay. My husband later told me the hitch was not properly seated and locked on the ball and could've bounced off at any time.

SAFE TRAILERING

There are books and websites that address safe trailering. As already quoted, Cherry Hill has written two wonderful books, *"Trailering Your Horse"* and *"Equipping Your Horse Farm"*, covering both trailers and training your horse to travel. There are plenty of pictures to demonstrate the text. http://www.horsekeeping.com, Cherry's website, has a multitude of free articles, as well. The *"Equispirit"* website, http://www.equispirit.com, has detailed descriptions of everything from laws regulating animal transport to how to choose a tow vehicle. Please be sure you're fully informed of all aspects of transport before you hit the road.

While driving on the highways, you often see trailers carrying horses already tacked up with saddles and bridles. It's very tempting to try this. The argument for traveling this way is that you're ready to hop on your horse once you get to your destination. Your horse may be somewhat anxious about being saddled in a strange place or be excited about the ride and will dance around, trying your patience. Your friends will be saddled and ready to go and you'll be chasing your horse with saddle in hand. You'll start your ride with a lathered horse, and you'll be out of sorts, and hot under the collar.

There are problems inherent in this method of travel unless your horse is trained to travel this way. If he rides with another similarly dressed horse, and, in moving around the stirrups lock together, one or both of the horses may become agitated and start to climb the walls. If one horse falls in the trailer, it will be difficult to right him and just as difficult to remove the tack while he's on the floor. If one horse is down, one of the other horses may become tangled in the equipment and become injured. If the tack becomes tangled, parts may break for which you won't have spares in your travel kit. Your ride will then be over before it starts.

And, from a rescuer's point of view, if a horse is down and he needs to be pulled out, the tack will be a time-consuming obstacle to overcome and will be cut free from the horse in whatever way is the quickest, with no regard for the tack.

Here's a sobering picture: A load of Civil War re-enactment horses travel in a stock trailer for approximately five hours. They've been on country roads, freeways, city streets, and through rush hour traffic. There are 7 horses crammed into the trailer, all tied on the left side. They are used to traveling this way; they do it every weekend during the summer months. The trailer arrives at its destination and, upon opening the door, the driver finds that one of the middle horses is down. Her head is under the front legs of one horse, her legs are between the legs of another horse. All horses are anxious to get out of the trailer to stretch their legs and eat. The downed horse isn't moving.

This is a true story and fortunately had a happy ending. The horses were all Standardbreds – ex race horses – and they had the typical Standardbred's calm disposition and common sense. The vet was called and the first horse was unloaded. The second horse was slowly led out from amongst the downed horse's legs. She picked her way carefully through the legs without stepping on them. The horses at the front were calmed by a bystander's voice. The downed horse was pulled out from under the next-in-line horse and out of the trailer. Upon hitting the ground, she stood up, shook herself and began grazing. If these horses WEREN'T Standardbreds there might have been a different and very ugly outcome. Yes, I'm biased toward this breed as you might have guessed.

One important consideration when driving on roadways: A study of more than 200 trailer incidents has shown a large number of them involved gooseneck trailers becoming stuck on train tracks at crossings with raised tracks. Watch for signs cautioning low clearance and take an alternate route. If you ARE caught on the tracks, call 911 immediately and unload your horses, pets and passengers. The call will alert railway companies to warn their approaching trains, and unloading your horses will keep them safe and also may lighten the load enough to complete your crossing. Make sure you take your horses to a safe area as far away from the tracks as possible. With a straight load trailer, make sure your jack stand is raised to its highest possible point.

SAFE DRIVING TIPS

Your horse should always wear a head protector and leg wraps when traveling

- If you are hauling one horse in a two-horse straight load, the horse rides on the left side. If you are slant-loading horses, they are tied on the left side. This is because most roads are "crowned", which means higher in the middle. Concentrating the weight on the left or "high" side keeps the trailer from wanting to tip over
- Make sure the hitch on your vehicle allows your trailer to travel level. If your horses are always standing uphill they may scramble or constantly shift position to try to compensate for the angle. At the very least, they will arrive at their destination stiff and sore and probably out of sorts
- Do not EVER load a horse facing backwards unless the trailer is specifically designed for rear-facing travel. This puts the majority of the weight on the back end of the trailer and excessive strain on the hitch

Photo courtesy of Dr. Rebecca Gimenez

- Under inflated tires on trailers and vehicles can cause sway so make sure your tires are properly inflated before each trip. Uneven wear on your tires can be caused by uneven inflation or by poor axle alignment in the trailer

- NEVER remove the bars off the windows so your horse can stick his head out while you're driving. Not only does this practice lead to serious eye problems, but if another vehicle gets too close, your horse's head could be injured

YOU CAN HELP

Finally, what should you carry in your truck or trailer to aid emergency responders? Remember, very few rescuers will have had the training necessary to work confidently and safely around your horses. Some rescuers will have a negative attitude about helping your horses. And, most rescuers will not have the specialized equipment that will make the job easier and safer.

The most important thing to carry is a good attitude. You can help on scene, but you will not be in charge. Be respectful of the chain of command of the people who are putting their lives on the line to help you. One of the biggest challenges at horse accidents is the owner!

It's quite possible that you will also be affected by a trailer accident. You may be physically injured or you may be emotionally unable to handle such a situation. You never know until you've gone through it. One way you can help is to be prepared.

THE BOY SCOUTS WERE RIGHT!

Since we were kids, we've heard about the Boy Scout's Motto: "Be Prepared". Along came the 60's and 70's and young people of that time decided it was "square" to be a Boy Scout, and paranoid to drag around stuff you weren't going to use right away. Now we're somewhere in between. We take our cell phones most everywhere we go, both for keeping in touch and for emergencies. We have our disaster kits ready at home as part of "Homeland Security". Now its time to prepare yourself for the road.

As mentioned above, rescuers will probably not be trained in technical horse rescue. The good news is that a lot of the equipment they will need is already in their emergency vehicles. They probably don't even realize it, but with a few added supplies from your trailer or truck, they'll have your horse out in no time at all!

What is it they carry that will prove so helpful? All fire engines carry hose. Hose can be used to drag a recumbent horse. This is used in a forward pull, where the ends will run between your horse's front legs, and in a backward pull, where the hose will go over your horse's back and the ends will run out between his back legs. This puts pressure on the boney structure of your horse, not his backbone.

Another useful tool is a "pike pole" or "ceiling pole" which has a hook on one end. This hook is used by firefighters to pull down ceilings and roofs in burning buildings. It can also be used to slide under or beside your horse to pull rope, webbing or hose into position.

Sometimes fire hose is too heavy and cumbersome to be effective. Then rescue personnel will use the webbing they carry. This is very strong and wide enough to avoid putting dangerous pressure on legs but thin enough to slide under a horse. The webbing can be tied to the fire hose to draw it under the horse. Or, it can be made into a short-term harness to lift a horse.

No one will expect you to carry fire hose or pike poles in your trailer! You can, however, purchase a roll of webbing. It has all kinds of practical uses in and around the trailer. Now your buckets can be tied up by webbing instead of baling twine! Buy lots and use it to make tie straps with quick release snaps, or to tie stirrups together for traveling. Telescoping boat hooks are also very useful, both as a substitute for a pike pole and for fetching objects from the dark recesses of your tack area.

Here's a list of items to carry with you for your own safety and that of your horse. Unlike the organization with a capital B and S, girls can be boy scouts too!

- Number one on your list is **identification**. You should be able to identify yourself and your horse. You'll have your driver's license but for your horse an I.D. sheet with pictures will help identify him. Go to www.redjeansink.com to download a Word document that allows you to add your pictures digitally. Make sure at least one of the pictures includes you. If something were to happen to you – you're taken to the hospital, for instance – and your horse is transported, it will be a lot easier to get him back if you can show an identification sheet with pictures. If your dog travels with you, make up one for him too! Make sure it's in a place where it can be found, such as your glove box or your traveling pet disaster kit. Also, please microchip your animals
- Next to identification, the most important paper you can carry is one detailing instructions of **who to contact** in case you are incapacitated. List friends who know you and your horse, and your veterinarian. Copy and fill out the forms in the back of this book provided by USRider. They are a "Limited Power of Attorney for Health Care" and "To Emergency Responders". These should be kept with your animal's kit and/or in the glove box of your vehicle
- Program your **emergency numbers** into your cell phone and designate them with the acronym ICE, which stands for In Case of Emergency. For instance, if your contact is your sister Sally, create an entry in your phone's address book that reads "ICE – Sally". Emergency responders are trained to look for this on victim's cell phones. If you use a roadside assistance program such as USRider, program their emergency contact number and your membership ID number as well. Keep all phone numbers current and advise the people on your list that you included them and what your wishes would be in the event of an accident

- A **cell phone** is a very handy means of communicating. Some areas are considered "black holes" – it's almost impossible to get a useful signal. But the freeway at night, or in the middle of a strange city, can be awfully scary places. If you have an accident with your trailer and you're stranded the cell phone may save your life or your horse's
- Your vehicle should always carry a **first aid kit**. Keep it filled with supplies for both animals and humans. In most cases, what suits one species will suit another
- If your horse is injured, when you call 9-1-1 tell the dispatcher to send a **large animal veterinarian**. The emergency responders could call when they arrive on scene, but that will use valuable time. Often a large animal vet will often take more than an hour to arrive. If you're close to home and uninjured, you can call your own vet and get instructions on what to do to help your horse until she arrives. Always tell the dispatcher you're traveling with animals, how many and their species, and that they will need to be accommodated, if necessary
- Carry a lightweight **blanket** and a container of **water**. If your horse is stressed or injured you will want to make sure he doesn't go into shock. A blanket will help keep him warm. If he's recumbent and thrashing, a second blanket or towel will protect his head and will keep his eyes out of the dirt. Anxiety may cause him to become very particular about the water he drinks. If he has some from home it may soothe him. Also carry an extra jacket for yourself. An accident will likely happen when you are underdressed, it's raining/snowing or dark/windy
- Carry an extra **lead and halter** in your truck. The ones on your horse may be destroyed; extras in the tack room of the trailer may be unavailable if the door is damaged or blocked

TRAVELLING WITH YOUR SICK HORSE

Written by Nicole Ehrentraut,
DaVinci Equine Emergency Transport

Travelling with a sick horse can be daunting at best. Whether your horse is 3 legged lame, colicking or literally lying down, you will need to carefully consider the following.

Before you set out ask yourself, IS TRAVEL NECESSARY? Will travelling to the vet hospital likely end up in my animal's recovery? This is something you have to decide together with your local veterinarian. It's important to take into account that any "rescue" effort has risks, and they have to be weighed carefully as to the best possible outcome for your horse. If you feel travel is necessary, here are some tips to make your trip safer and more comfortable for your horse.

Items to Travel With

- **Video Camera**

Cameras can prevent incidents from getting out of hand. For example, your horse gets his hoof stuck in the hay net. You can pull over before he becomes so stressed that he falls or ends up jumping over the divider. If you have an injured horse, you can tell how well he is travelling, how slow you should be taking curves, is he about to fall, etc.

- **Fly Mask & Human Life Vest**

If your horse is down, you need to protect his head. The fly mask with protect the eyes from scratches and the life vest can be put on around the head to protect the head from injuries.

- **Horse Helmet & Leg Wraps**

You can buy a small helmet from a tack shop which protects the horse's poll. This should be worn at all times. Studies show that horses with fatal poll injuries during trailer incidents may have survived if they had had worn a helmet. Use leg wraps. I'm not a huge fan of the big shipping wraps. Most horses are not used to them and they can hinder rather than help. Use exercise boots or polo wraps to protect your horse from injuring the lower legs where the bone and skin is so thin.

- **Fans**

Especially if it is hot. Better to use fans than have open top doors. Debris (including lit cigarettes) can come in through those open doors and injure your horse, and if your horse is already stressed due to an injury, he may panic and try to escape out of anything that is open.

- **Inner Tubes**

ATV tire sized inner tubes are great to hang along the side of your trailer if your horse is 3 legged lame and will wobble around a bit. Load your horse to the left of the divider in a straight load and "package" him in with the inner tubes. You don't want it to be too tight; you just want the horse to have something soft to lean on. As with anything new, monitor your horse with this new item BEFORE driving anywhere to make sure he is comfortable with it.

- **Safe Travel Documents**

ALWAYS travel with a sheet that identifies your horse and any ailments he may have and/or medications taken regularly.

Power of Attorney Sheet – who has the legal right to make decision for your horse if you can't?

Medical Release – how much are you willing to spend on surgery for your horse and when is it ok to euthanize? If you are not conscious for whatever reason, the attending vet and first responders will appreciate knowing your wishes.

- **First Aid Kit/Water/Hay**

HUMAN & HORSE. You never know who will need it. If you are travelling with an injured horse, he will be more likely to injure you.

TIPS FOR DRIVING WITH AN INJURED HORSE

- Drive like you have a full cup of coffee on your dash that you don't want to spill. Your horse is less likely to be able to balance himself well, so drive slow and steady. Even more so than you usually do. Use that video camera to monitor your driving and the horse's response. Adjust accordingly.
- If your horse is 3 legged lame, it may be better to put him in the left side of a straight load trailer with inner tubes for support. That way he can lean up against the side.
- If your horse is colicking and is likely to fall or try to lie down, remove any dividers and allow the horse movement. If he falls/lays down and is calm, continue travelling. If he is floundering around, make sure you put head protection on him. If it is too bad, you may need a vet to sedate during travel.
- NEVER TRAVEL IN THE BACK WITH YOUR HORSE! You will be in danger the entire time and may put your horse in danger too. There is nothing you can do in the back to help your horse that is worth the life risk.
- NEVER TRAVEL ALONE WITH A SICK HORSE. I know this is impossible at times, but try not to. If something happens, having an extra person there is much, much safer.
- Have a list of vets along the way in case you need one during a longer drive: horseforum.com/equine-vets
- Think several steps ahead at all times. The key to a safe trip is asking yourself all kinds of "what if" questions. First Responders always have a Plan B & C waiting in the wings just in case.
- Use your common sense. If you feel like you are making a decision based on emotions, step back for a minute. Ask someone else who is not as emotionally attached as you are to

help. Also, your horse will mirror your inner and outer behavior. It's hard to keep calm, and to drive slow, when all you want to do is speed down the road and get there. TRY…

Thanks, Nicole

YOUR FIRST AID KIT

The following items, kept in a watertight container, are considered a "basic first aid kit". Keep it handy in your truck.

- Nolvasan or other disinfecting solution
- Non-steroidal sterile eye wash
- Non-steroidal antibiotic ophthalmic (eye) ointment
- Triple antibiotic ointment
- Anti-bacterial spray (for large surface wounds)
- Gauze rolls, adhesive tape, vet wrap
- Sterile pads – 2x2, 4x4, non-stick (Telfa) wound pads
- Sanitary napkins or cloth diapers or roll of cotton for padding
- "Standing bandages" – 2 quilted or fleece lined
- Duct tape (one of the most valuable tools in your kit!)
- Forceps, bandage scissors, tweezers, pocket knife
- Wire cutter, hoof pick, needle nosed pliers
- Sponge for washing wounds
- Vicks Vaporub, to dull your horse's sense of smell
- Flashlight on headband and a Megawatt flashlight
- Homemade ear plugs – stuff cotton balls in the toes of pantyhose or tie tampons together with string

Also good to have on hand are:

- CLEAN WATER to wash wounds (at least 5 gallons)
- Stethoscope, digital thermometer, bulb syringe, eyedropper
- A turkey baster – plastic – if you give drugs such as "bute", and some way to grind the tablets
- Abdominal pads for large wounds
- Q-tips and cotton balls
- Splint material in various sizes (broomsticks, 1x2 lumber, or PVC pipe sections, split lengthwise)
- Insect spray – spray and roll-on
- Hacksaw and a Twitch
- Towels, extra blanket
- Chemical hot and cold packs
- SWAT (thick pink ointment with fly deterrent in it)
- Extra cotton lead ropes and halters, collar and leash for your dog
- 100' cotton rope or webbing
- Heavy plastic gloves, heavy work gloves
- Tarp, trash bags, a clean bucket, and a shovel
- Notepaper and pens, "Stickies", something for fastening notes
- More duct tape! And a permanent ink pen ("Sharpie")
- Electrolyte powder, Gatorade or Pedialyte
- Battery operated clippers
- Closed-cell backpacker's pad ("Ensolite") for splints
- A reliable watch with an easy to read second hand
- I.D. collars and permanent markers for identification

YOUR RIG'S FIRST AID KIT

Just as you will carry supplies to care for you and your critters, your trailer and vehicle may need some help.

Here is a sample list of items to keep on hand in your rig.

- Extra cash
- Tool box
- Extra bulbs and fuses
- A wiring repair kit
- Fire extinguisher
- Spare tire and the equipment for changing the tire
- Spare belts and hoses for your vehicle
- Wheel chocks to keep your trailer stable
- Emergency flares or triangles
- WD-40 and duct tape
- A powerful flashlight – preferably one in the megawatts of power

Trailers and hauling vehicles never come with enough reflectors or reflective tape to keep them from being rear-ended. When you are parked along a dark road at night you're vulnerable, especially if the road is not straight and flat. When viewed by approaching vehicles at night you should "light up like a Christmas tree". Add extra lights if possible but also reflectors and reflective tape to all sides of both vehicle and trailer.

BUILDING YOUR OWN "GO BAG"

There are many good sources of information on building your own "go bag". Here are some suggestions to get you started. A duffle bag will hold these items, or you may want to use a backpack which is easier to carry.

You may want to purchase specialized equipment such as Rescue Straps for your bag. Please check the Commercially Available Equipment section in the APPENDIX. The top list contains generic equipment you can use to start, and the bottom list shows the typical contents of a commercially made kit.

Gloves
Helmets
2 60 Ft, 2- 3 inch two-ply webbing
2 25 Ft, 1 inch webbing
2 100 Ft, half inch "kernmantle" rope
Smaller ropes, various sizes and lengths
Pulley to use with the half inch rope
Boat hook
Cordless reciprocating saw and charged batteries
Duct tape
Horse blanket
Lengths of baling twine
First aid kit

Kernmantle ropes are constructed with an interior core (the kern) protected with a woven exterior sheath (mantle) that is designed to optimize strength, durability, and flexibility. Ropes with kernmantle construction are used in climbing, caving, and sailing. Prusik loops are typically made with kernmantle rope.

OWNER'S TRAVEL KIT CONTENTS:
Pack
1 3 inch Steel O-ring
1 Knife
1 Paramedic Shears 7 ¼ inch
1 LED flashlight with photo-luminescent shroud
1 CMC Rope Rescue Field Guide
1 4 inch Wide Rescue Strap - 18 Ft
5 2 inch Steel Pulleys
8 Large Steel Locking D Carabiners
1 50 Ft Rescue Line ½ inch
1 30 Ft Rescue Line ½ inch
1 25 Ft Waterline 3/8 inch
1 1 inch Tubular Webbing 5 Ft
1 1 inch Tubular Webbing 12 Ft
2 1 inch Tubular Webbing 15 Ft
2 1 inch Tubular Webbing 20 Ft
1 1 inch Tubular Webbing 60 Ft
4 Prusik Loops 5 Ft
4 Prusik Loops 6 Ft

This page intentionally blank

KNOTS

Students in a LAR class learn to tie knots

There are hundreds of knots, and a well-rounded Boy Scout, rock climber, sailor or firefighter knows many of them. But, in case you don't have a boy scout handy, and the fire department hasn't arrived, here are a few basic knots. A word about terminology: There are three parts to a rope – the standing part, the running part (the active end of a line used in making the knot), and the fall (or "bitter end"). Here's a simple way to remember these parts. One part is standing around, not being involved in tying knots, the other is running around itself becoming a knot. The fall is the tail end of the rope, the part that falls away from the knot. It's bitter about not being used. A bight is a curved section that does not cross over itself to become a loop.

CAUTION: THESE KNOTS ARE INCLUDED FOR YOUR INFORMATION ONLY. They are NOT included so you will try to effect a trailer extrication or other maneuver that is within the scope of training of a emergency responder. Leave the dangerous work to the professionals.

Why list knots, then? You can use them to tie up shelters; help an animal keep his head above water until help arrives; haul your gear or tie up a rescued horse. You'll find uses for them. Just, please! Be "Safe and Sane".

A knotman's axiom: If one knot is good, two are better.

BOWLINE – The "King of Knots". It can be used to tie objects together and for hauling. If properly made will not slip or jam and is easy to tie and untie. By making the loops large you can use as a sling, or you can hook the "eyes" onto a winch. Make a small loop. The "standing part" should be underneath. Put the "running part" through the loop and run it around the "standing part". Bring it back through the loop. Tighten by pulling on the "standing part" while holding the loop.

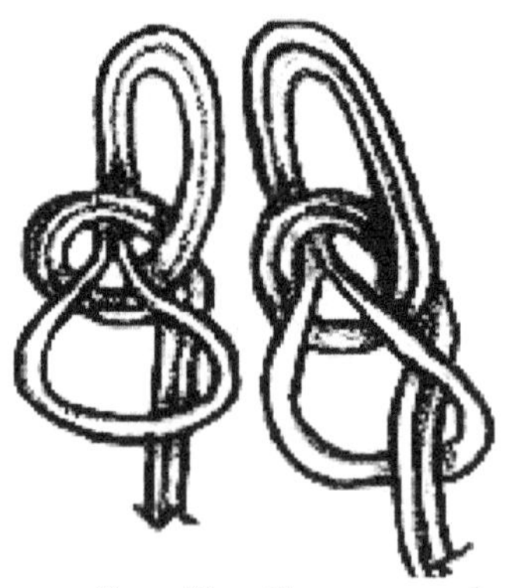

<u>BOWLINE on a Bight</u> – If you tie your bowline on a "bight" (loop) you can make a sling too. Make a loop at the end of your rope. Holding both pieces together, make an overhand loop and run the "bight" back through it. This will make a loop where the "bight" was brought back (called the "eye"). Open up the "bight" and take the "bight" back over the overhand loop, the "eye" and back past the overhand loop. Pull on the "standing and running parts" with one hand and the "eye" with the other until the knot is tight.

<u>DOUBLE HALF HITCH</u> – Use this knot for attaching a rope to a pole. Push the loops together and tighten by pulling on the standing part. The "running part" goes around the object. It crosses over the "standing part" and loops through. When the "running part" comes out of the loop, cross over the "standing part" again and bring the "running part" through the second loop. Snug up the loops.

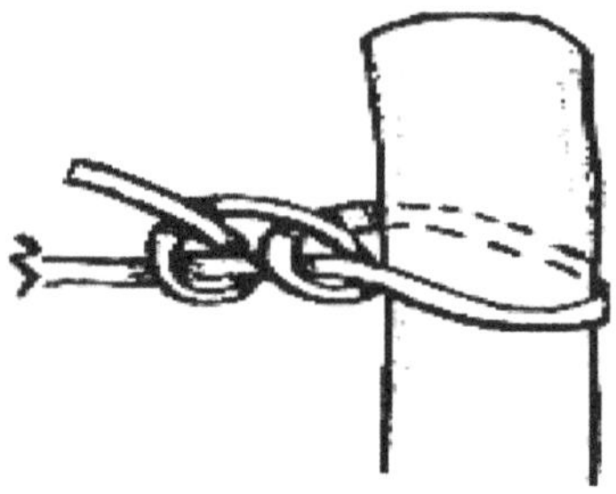

<u>CLOVE HITCH</u> – This hitch puts little strain on the fibers of the rope. While easy to tie and untie, it is not secure. Stand facing the post (or wherever you are planning on using this hitch). Wrap rope around object with the "standing part" on top. Wrap the rope around the object again. Feed the "fall" through the loop formed by this second wrap. Pull tight.

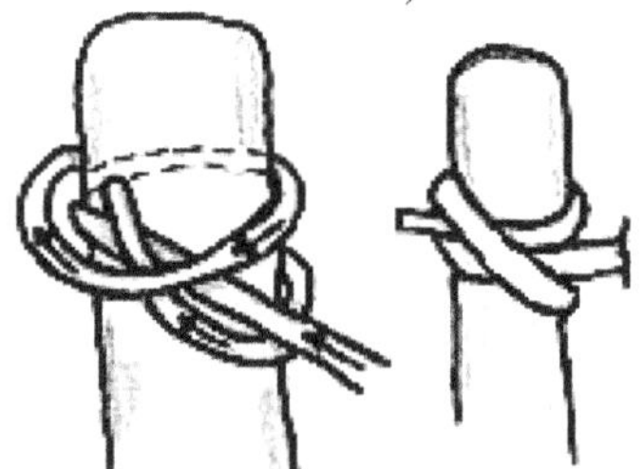

SHEET BEND or BECKET – A great knot for joining two ropes together. Works well with different sized ropes. Make a loop at the end of the heavier of the two ropes. Run the second rope up through the loop, over the two ends of the loop and underneath itself where it comes out of the loop. Tighten by holding both pieces of heavy rope in one hand and both pieces of lighter rope in the other. Pull your hands apart.

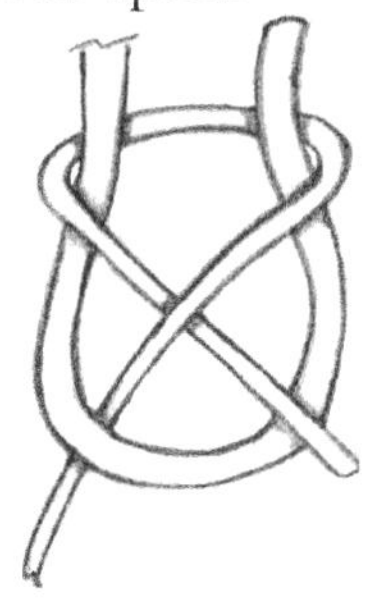

SHEEPSHANK – Use this knot to take up slack in or shorten a rope that is already fastened at both ends. If used without tying half hitches in the ends it can fall apart. This is a valuable knot to use if your rope has a weak point but is otherwise useful. Tying this knot over the weak point will bypass it. Make an "S" shape with the slack of the rope. Make a loop at one end and slide it over the "bight" of the "S". Make a loop at the other end of the "S" and slide it over the "bight" of the "S". Leave enough of the "bight" of the "S" to use it to make a half hitch, if possible.

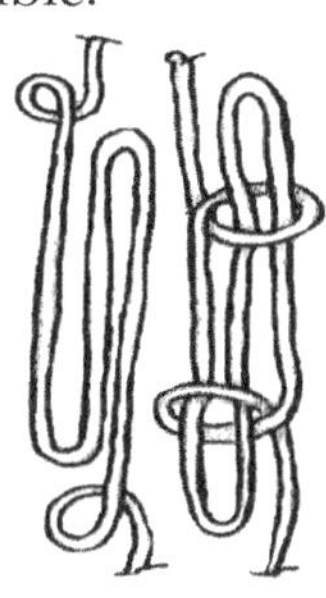

PRUSIK – If you need to attach something on the length of a rope, this one can be used. It's a mountain climbing knot. Use a short piece of rope, knotted at the ends to form a loop. Place a "bight" of the loop across the rope. Pass the other end of the loop behind the rope and through the "bight". Pass it through again so there are two wraps of the loop on the rope. Push loops together and pull tight.

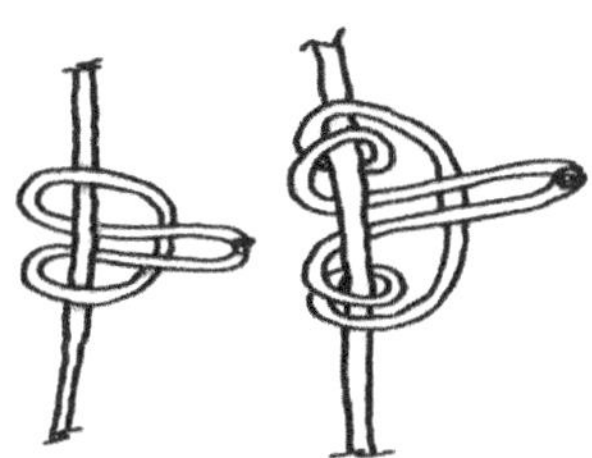

QUICK RELEASE KNOT –Run the "bitter end" over the object. Twist it to form a loop with the end underneath. Form a bend in the end rope and pass it through the twist and push down to the object. You can release this knot by pulling on the end. Good for tying horse to something.

TRUCKER'S HITCH – Use this hitch to cinch down a load. Can also be made on a "bight" if the load is secured at both ends. Form an overhand loop in the "standing part" and push a loop through it. Pull this knot tight. Feed the "bitter end" of the "running part" through or around the attachment point – post, tree, hook, etc. – and up through the loop at the knot. Cinch down and tie off with a couple of half hitches. If your rope is secure at both ends but needs to be made tighter, make a loop and tie it near one end. Feed line through the loop and cinch down. Tie off with a couple of half hitches.

WATER KNOT –Two overhand knots running in opposite directions. After you tie the first overhand knot, the second strap or rope is threaded along the knot in the reverse direction. The knot should be arranged neatly and pulled tight. Several inches of the strap/rope should run past the knot.

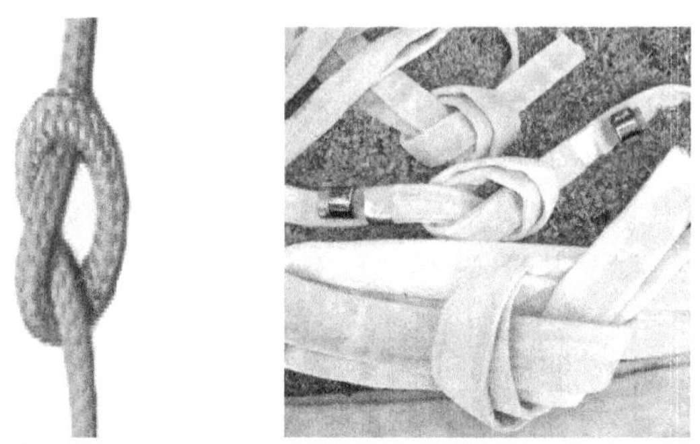

Great for web straps
Photos courtesy of Vicki Schmidt

GIRTH HITCH -- This simple hitch is used to attach a loop to an object. It can be used for hanging gear as well as tying into the middle of a rope. It's also known as a ring hitch and a lark's foot. Take a loop of rope, create a bight, pass that behind your object, take the other end and pass through the bight, pull tight.

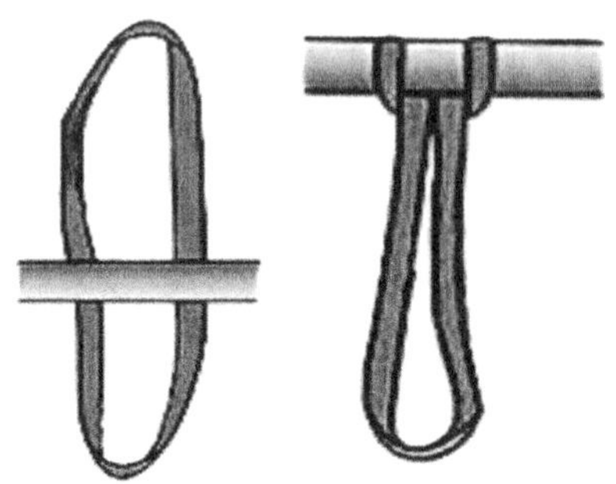

CONCLUSION

And there you have it. In this book we have provided some important information for the emergency responders who may be called upon to help save your horse. With this information they will be able to extricate your horse from a trailer, lift him out of a hole, or assist him up a slope --- pretty much any dangerous situation in which your horse will find himself. Responders will have the necessary information to keep themselves safe around large animals, whether they are leading your horse to safety or helping the veterinarian keep a downed horse from struggling to his feet. We've provided information on obtaining the tools of the trade and, most important of all, training in this specialized field.

For you, the horse owner, we have provided a wealth of information about horse trailers from the aspect of the safety of both you and your horse.

This book will not prevent accidents from happening. Also, it is not meant to replace professional training or professional help. But, hopefully, it will bring you some peace of mind.

Happy Trails!

This page intentionally blank

APPENDIX

This page intentionally blank

TIPS ON HANDLING LIVESTOCK

The following are a few key pointers on how to respond to an incident involving livestock and how to handle livestock other than horses. There is so much more involved in dealing with each species the information would fill a book on its own. I highly recommend pursuing some training in this specialized field. This section was written by Jennifer Woods, who teaches classes in Livestock Emergency Response. The first part of this Chapter outlines specialized information by species. The second part will give you a general outline of what is required at accident scenes involving commercial livestock trailers. The information is a very basic guideline as these accidents are quite volatile and challenging.

The Livestock Emergency Response training program will provide you with the skills and knowledge required to respond to livestock transport accidents in a way that is safe and efficient. The curriculum includes handling and behavior of all domestic livestock, especially in times of stress; trailer design and structure, extrication from the different compartments of commercial livestock transports; what tools to use and not to use; where and how to cut; removing animals from trailers; capturing and containing loose livestock; and, euthanasia procedures. Jennifer's contact information is at the end of this chapter, as well as in the RESOURCE section.

CATTLE

- Cattle will stampede if frightened. Their instinctive reaction to danger is to flee
- Cattle will calm quicker if left in a herd
- Cattle will kick or charge when frightened

PIGS

- Pigs are difficult to drive; they cannot be chased
- Running pigs may kill them
- Pigs are sensitive to heat and cold. They can easily suffocate in heat and should be dealt with quickly
- Pigs can be cooled by misting them with cold water. Do NOT pour cold water on them as the shock may kill them
- Pigs are sensitive to frostbite; wind chill can kill them. They MUST be blocked from the wind
- Pigs do not like to step up or down so ramps must be used

SHEEP

- Sheep will instinctively bunch up in a tight herd, causing suffocation
- Do not lift sheep by their horns or fleece

LLAMAS/ALPACAS

- Llamas are extremely heat sensitive. They need to be removed immediately from a stalled or rolled trailer in the summer
- Llamas will lie down when they are stressed. They can be dragged by their necks
- Llamas do not like to be petted. They WILL bite
- Do not blindfold a llama. It will cause him to freeze on the spot

ELK

- Elk are very quick and agile. Cannot be contained by typical fences
- Elk are easily baited with food. Use this trick to move them
- Elk respect height. Use your tallest people to deal with them
- Call in an experienced handler

BISON

- Bison may gore you when they feel threatened
- Bison do everything at a run. They cannot be stopped
- Call in an experienced handler

POULTRY/TURKEYS/CHICKENS

- Poultry are frightened by close contact with people
- If crates or cages have spilled and birds are still inside they should be uprighted immediately. The birds will suffocate very quickly
- Birds are easily affected by cold and heat. They need to be protected from the elements
- Do not chase birds or cause them to fly. When handled calmly they can be herded

OSTRICH/EMU

- Ostriches can be extremely dangerous. They have a powerful kick and sharp talons. They will also peck with their beaks
- Ostriches are very fragile. They break easily
- Ostriches panic very easily

INCIDENCE RESPONSE FOR LIVESTOCK TRANSPORTS

When accidents happen involving commercial livestock trailers, the scene is often dangerous and confusing.

There is an incident within an incident occurring. There is a motor vehicle accident, and there is a livestock incident.

Most emergency responders have little or no experience in handling livestock, and most livestock handlers have limited experience or training in rescue and recovery and in the handling of stressed livestock.

By being prepared for an accident before it happens, and understanding how to effectively respond to an incident involving livestock, the welfare and the safety of the emergency responders, the handlers, and the animals will be greatly enhanced.

There are also financial benefits to a coordinated response. When an accident is handled by a knowledgeable response team, the economical losses can be significantly lessened. Fewer animals may need to be destroyed, the structure of the trailer may be salvaged when cut properly, and the cost of the recovery, rescue and roundup will be less.

Responding to an incident

- Call 9-1-1. Advise dispatch of the accident location, type of livestock on-board, number of animals, if there are any loose livestock, and the need for a livestock veterinarian. Advise that police and fire approach with their sirens off, if possible. Advise dispatch to contact animal control, if that procedure is mandated in your area
- Herd any loose livestock from the road. Gather them in one area, as far away from traffic as possible
- Light flares and set out warning sign to alert approaching traffic of the accident
- If possible, get the following information from the driver:
  - Species of animals
  - Number of animals
  - Are there any loose animals
  - Are there any hazards

Salvage, rescue and recovery

- 90% of all rollovers tip to the RIGHT hand side
- Extrication procedures differ vastly for a commercial trailer that rolls on the left side vs. the right side

- There are different procedures for "fat" trailers vs. standard commercial livestock trailers
- Fire departments are in charge of cutting the trailer open
- Never tear a trailer apart with a tow truck or winch
- Never enter a rolled trailer loaded with animals
- Never upright a loaded trailer
- If the trailer is sealed for international transport, the USDA must be notified

The laws and regulations pertaining to the loading of compromised animals apply to accidents

- Request the services of a deadstock removal company, if necessary

Submitted by:
Jennifer Woods
J Woods Livestock Handling Services
www.livestockhandling.net
livestockhandling@mac.com

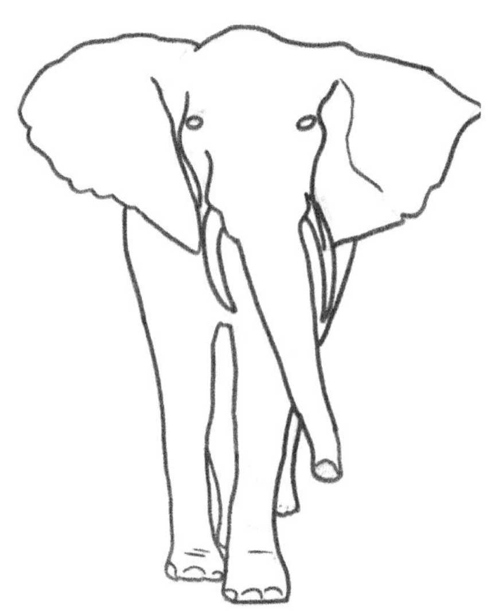

BARN FIRE SAFETY

Photo from the internet

FOR THE RESPONDERS

The following information is aimed toward emergency responders and will provide local fire and rescue personnel with the knowledge and tools they need to plan, prepare and safely respond to horse barn/stable fires.

RESPONSE & ARRIVAL

Horses have heightened senses that far exceed those of humans thus they see, hear and smell more intensely than us. Therefore, high pitched, shrill, continuous sounds can aggravate a horse and the bright flickering of multiple strobe lights can cause additional stress. When responding to barn fire calls either make a silent approach or cut the sirens several blocks before the address. Once on-site, cut all flashing lights.

When arriving on scene, cordon off all egress, such as driveways, footpaths etc. This can be accomplished by closing existing gates, positioning vehicles across openings and in some cases, using brightly colored caution tape to turn back the horses. Given the opportunity, a horse could run out one of these natural 'funnels' and onto the road thus posing a public safety threat.

If horses are loose on-site when you arrive, every attempt should be made to move them to safety. Loose horses pose an enormous threat to the firefighter, the public and to themselves.

Set up a command post, locate the person in charge of the stables, and determine the following:

- How many horses are in the building?
- Where are they located?
- Are there any danger zones such as dead end or unmaintained roads, mud holes, septics, and places in the stable where combustibles are stored? (Hay, fuel, tack cleaner, bedding)
- Are there additional accessible water sources such as tanks or ponds? Are they compatible with your equipment? (standpipes, couplers)
- Where are the shut off locations for utilities?
- Where are the safe containment areas (paddocks, pastures, pens)?
- Layout of the building (such as outside stall doors, tack rooms with combustibles, stalls as hay/bedding storage, and other exits)
- How are the stall doors latched, which way do they open, and are there kickboards across the entrances to stalls?
- Who is experienced in managing horses and can help?

Remove all SUV's **"SELF DEPLOYED, UNINVITED VOLUNTEERS"** from the site. ***They will do more harm than good.***

HORSE HANDLING

During an emotionally charged event such as a fire, expect that horses directly involved in the fire, whether in a building or in a paddock with close proximity to a burning barn, will be visibly upset, and may attempt to get away from the fire location.

Approaching horses during times of stress should always be done in a calm, quiet manner. Relax your arms by your side and relax your body to show the horse that you are not the aggressor. A large breath in and out will help.

If possible, approach the horse from the left side and if you are able, stand at the horse's shoulder. A comforting scratch on the shoulder, gentle rub and a soft voice will help to calm the nervous horse. Never slap or vigorously pat a horse; this is perceived to be an aggressive move and will only heighten the horse's fear.

Once you have gained the horse's trust, you should be able to slip a halter over his head or tie a rope around his neck.

DO NOT stand in front of the horse or directly behind the horse. Both are considered 'Danger Zones'.

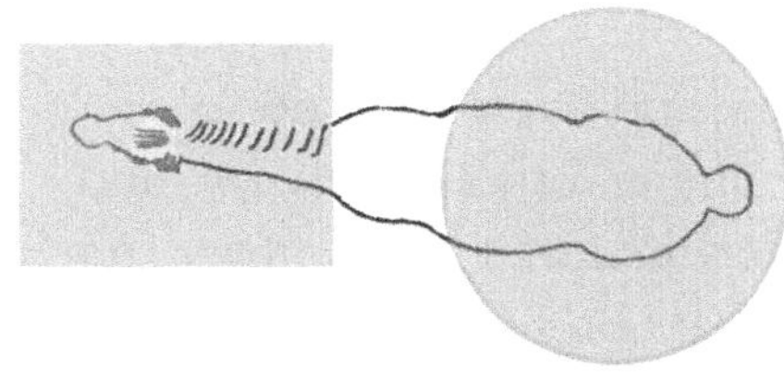

REMOVING THE HORSE FROM THE BARN

If circumstances allow for horses to be removed from the barn always lead the horses closest to the exits out first and make your way down the aisle removing horses as you work your way in and closing stall doors behind you.

Do not attempt to remove horses alone. Always use a 'buddy system' when doing stall releases. Have one firefighter manning the stall door, another manning the barn door and a firefighter haltering and leading the horse out of the building. Like little children hiding in their rooms in a house fire, horses frequently attempt to return to a burning barn because it is their 'safe place'.

Ideally use a halter and lead rope. Some barns will have halters and lead rope hanging at every stall door, others will keep them all together at the entrance to the barn or in an equipment room. If you can't find halters, tie an emergency halter (SEE APPENDIX: Rescue Equipment).

Never wrap a lead rope around your hand. Should the horse bolt you will be dragged and could conceivably break your fingers/hand. Instead, fold the rope in the palm of your hand so that if necessary, you could drop the rope and not be entangled.

If halters and lead ropes are not readily available, a rope around the horse's neck secured with a loop can be used. Other items such as belts, webbing, and narrow fire hose can also be used to lead a horse. Loops around the neck should be used only when a horse is mobile and moving willingly, and you should never tie a horse using these makeshift halters.

NEVER use the horses head, neck, legs or tail to pull him to safety if he is down. This could cause death or serious, potentially fatal injury.

Never turn horses loose as a way to evacuate a barn. This poses some real safety issues not just for the horse but for the firefighters and other emergency personnel not to mention the public safety threat should horses end up on a main road. Instead, make sure all of your firefighters know the basics of emergency horse handling and emergency stall release.

Do not blindfold horses unless you have been trained to do this. Some horses react negatively to being blindfolded and it is probable that a blindfolded, distressed horse will be too much for the average untrained responder to handle.

If you have identified qualified horse handlers they can assist with horse handling and leading animals to safety.

The majority of barn fires occur in the winter, and in the early morning hours when darkness makes fighting a fire doubly difficult. We offer the following suggestions for preparing your firefighters for the inevitable barn fire response:

- Large amounts of hay may be stored overhead in hay lofts and that hay will burn fast and hot. Therefore, always establish where hay is stored in relationship to the fire before you decide to enter a barn fire situation

- If there is a loft fire, close all 'hay drop' doors, if any. This will lessen the draft effect and shut off additional oxygen

- If entering a smoke filled barn, be prepared for obstacles in the aisles – saddle racks, tack trunks, stacked hay, buckets, blanket racks or hoses. All could trip you

- Take great care while navigating in old converted dairy barns. Old dairies probably won't have a center aisle, instead being a labyrinth of stalls, ramps and walkways on different levels throughout the barn. Some have abandoned manure pits under the main floors

- Tack and grain rooms are great ignition sources. Tack rooms contain leather saddles, cotton blankets, liquid cleaners, chemicals, and veterinary preparations. In grain rooms, virtually everything is combustible and could be a huge risk to firefighters

- Stall doors – some are sliding doors, and a few have doors opening in, others open out, and some are Dutch half doors, often with bars on the top. Latches may vary from sliding latches to flip latches and everything in between, and some stalls are latched top and bottom

- Many stalls have a "kick board" in the front of each stall on the floor to keep bedding in the stall. Be sure to check for "kick boards" to ensure you don't trip

- Some stalls don't have a door but instead use a "stall guard" – a grid of nylon webbing that hooks on either side of the stall opening and that will have to be unhooked to gain access. Nylon burn very hot; take great care handling them to avoid burning your hands

CONTAINMENT

If you are able to safely remove horses from a barn fire…you have to put them somewhere. Keep these suggestions in mind:

- Ask the person in charge the safest place to put the horses. An adjacent round pen, paddock or pasture removed from the fire is ideal

- Appoint a qualified person to secure all gates and monitor the "holding areas" until the fire is over. Emotional horses may either charge the gates or jump fences in an effort to flee from the perceived danger

- Unaffected buildings or indoor riding rings are good choices if the horses are able to be tied. It is not a good idea to allow horses free movement in buildings such as barns etc. as the propensity for injury is high. DO NOT LEAVE HORSES UNSUPERVISED IF THEY ARE TIED UP

- As a reminder – always cordon off all exits to eliminate the possibility of loose horses on the roads

TO CATCH A HORSE

Perhaps the most difficult task when coming upon a loose horse is catching him. Should a horse escape and gain access to public roads the task becomes inherently more difficult.

- Do not chase the horse with a vehicle. That will only heighten the horse's fear, make the situation worse, and you won't catch him

- Stop all traffic in the area to avoid an accident. A loose horse is a public safety risk

- Horses are "herd" animals, so introducing a calm horse (known as a "Judas" horse) could make the loose horse more comfortable, giving you a better chance of coaxing him close enough to be caught

- If the horse is not running in fear, food bribery is usually successful. A bucket of grain or bag of carrots is very effective

- If the horse is fleeing, go several blocks "downwind" of the horse and set up a road block with plastic fencing manned by humans. This MAY convince the horse to stop OR the horse could run around it, jump it or try to go through it. He may not even see you there. *Always attempt this measure with great caution*!

The bottom line is that horses will eventually stop when they are tired of running. You may find them munching on a patch of healthy green grass. At that point, attempt to approach the horse as discussed in previous sections. If you do not have a halter, a belt, webbing, or a piece of rope can be slipped around the horse's neck.

Until help arrives establish a makeshift paddock by either placing vehicles or trucks in a circle or perhaps locate a nearby fenced area where the horse can be moved.

Never use the legs, neck, tail, or head to catch a horse. The risk of injury to you is high; the risk of serious injury or death to the horse is higher.

FOR THE HORSE OWNER

The following information is aimed toward horse owners. Knowing what you can do to make your barn safer and hopefully prevent a fire will help you sleep better at night.

Although hundreds of horses – and thousands (even tens of thousands) of other confined animals – die in fires every year, 85% of all barn fires are the result of human negligence and very few facilities have fire prevention plans.

This checklist is designed to help you pinpoint the areas that need to be worked on and those that are already safe. Please do a walkthrough with your fire department and have an electrician evaluate your facility.

↔↔↔↔

- ☐ My address is posted clearly and visibly in large reflective numbers at the road
- ☐ My driveway/road can handle height and width of fire vehicles, there are no overhanging branches, and the roadway is gravel or other improved base
- ☐ I have dedicated turning access for fire vehicles and no one is allowed to park in front of the stable
- ☐ My fire department knows the location of all water sources on my property – ponds, wells, storage tanks, hydrants – and there is easy and safe access to them
- ☐ If there is a pond on my property I have installed a standpipe (a rigid vertical pipe to which fire hoses can be connected)
- ☐ My fire department knows the location of all emergency utility shutoffs
- ☐ If there is a lock on my gate I have a lock box and my emergency responders have access. If there is a combination lock, they have the combination
- ☐ There is a layout of the property in the lock box and I have given a copy to the fire department
- ☐ I have cleared a fire barrier along my road and my driveway, and all debris from around my stable
- ☐ There are no overhanging trees at my stable
- ☐ I have clearly marked and easily accessed water spigots on all sides of my stable, with attached hoses the full length of the side
- ☐ I have a shovel and a ladder on each side of my stable
- ☐ I have posted "No Smoking" signs around my stable and have informed my boarders that smoking will not be tolerated anywhere on my property

INSIDE

- ☐ I have a sprinkler system inside my stable. This system is appropriate to my climate and water system (even rural!)

- ☐ I have clearly marked utility shutoffs. My boarders know where they are and how to turn them off
- ☐ I have large fire extinguishers at each exit, and if my stable is large I have one hung every 50 feet
- ☐ Everyone at my stable knows how to use a fire extinguisher
- ☐ I have more than one doorway from the outside into my stable. Exterior doors are never locked
- ☐ Access doors open to the full width of the aisles
- ☐ Every stall has a door that opens to the outside
- ☐ If needed, equipment is in place outside the exterior stall doors that can be made into a safe runway to a paddock or pasture. When evacuating horses, exterior doors open against the outside wall and horses are herded down the runway to a safe holding area at least 100 feet from the stable. As each horse exits his stall, the door is closed behind him
- ☐ I have all electrical wire encased in non-corrosive conduit and all light fixtures encased in safety cages
- ☐ Cobwebs, hay and debris are cleaned up every day
- ☐ All doors to stalls are in good working condition
- ☐ I do not have any household extension cords, fans, or heaters in my stable. All equipment is designed for use in a stable
- ☐ Any household electrical appliances such as coffee makers use covered, grounded plugs and are unplugged when not attended
- ☐ I have a phone in my stable. It is easily accessible, clearly marked, and emergency numbers and directions to the property are posted close by
- ☐ Hay and bedding, fuel and vehicles that run on it, are stored in a separate building at least 50 feet away from my stable
- ☐ All aisles are clear of any hazards, including hay bales, tack boxes, electrical cords, and cleaning equipment such as rakes
- ☐ I have lightning protection on top of my stable
- ☐ I have used flame retardant paints on any wood in my stable
- ☐ Every horse has a halter and lead rope on his (inside) door; every boarder and her horse have practiced emergency evacuation procedures (horses have practiced with and without a blindfold); and every horse has been introduced to a firefighter in turnouts
- ☐ Every horse knows how to load into a trailer under any condition (bad weather, day/night, wildfires, veterinary emergency)
- ☐ Our stable practices fire drills and every boarder understands that

once flames are seen no one will be allowed to enter the stable, regardless of what **person or animal** is in the stable

☐ Safety and evacuation procedures are discussed at every boarder's meeting and everyone is aware of the dangers

Stable fires are the scariest of all scenarios involving your horse. Fire is so fluid, and moves like lightning. Bad things can happen in an instant to you, your horse, your property, your neighbors, and the people who come to help. Fire is capricious; it is not forgiving; it can even generate its own weather. Whether we are talking about wildland fires or structure fires, there are things you should know to keep yourself and your horse safe.

When you call 911 – from OUTSIDE the building – make sure you identify your exact location including identifying landmarks. Tell the dispatcher that live animals are involved and be specific that this is a STABLE fire not a BARN fire. Firefighters will typically respond with more speed and take more chances when they know lives are in danger.

FIRE FACTS

Research indicates that **95%** of fires in barns with sprinkler systems are extinguished. Sprinkler systems come in two varieties: wet – for mild climates and dry – for harsher, colder climates. There are also systems for rural areas with low water pressure.

Research has shown that it typically takes one minute to halter a horse and lead him 100 feet. Instead of working with horses in a smoke-filled stable, install exit doors to each stall on the outside of the building.

Although it goes against all logic, a horse will return to his stall in a burning stable. There is reason for this. If he is regularly stabled, he perceives his stall as a safe place. When he is scared witless, this is where he will return to find that feeling of security. CLOSE EACH STALL DOOR AS THE HORSE EXITS.

In the 1970s, a group of agencies studied stable fires and the Fire Protection Association published a report. What they found was:

- Horses can survive fire less than one foot in diameter or with a temperature less than 150°F at a level of 15 feet
- If the fire starts in a stall, the horse has less than 30 seconds to be rescued. Sadly, most horses in a burning stall will die from smoke inhalation
- In one to two minutes a burning bed of straw will generate more heat than a pool of burning gasoline
- While it takes about six minutes to engulf a stable in flames it takes about half that time to fill it with smoke

- Prey animals don't recover well from burns and often die days later from smoke inhalation, so be sure to have your vet check your horse immediately
- Once your horse is out of the fire zone, hose him off. Embers can get caught in manes/ tails, under blankets; nylon blankets and halters can melt onto your horse
- The **top three causes of barn fires** are: improperly cured hay (store it in its own building); electrical (run all wires in non-corrosive conduit); and carelessness (ban all smoking)

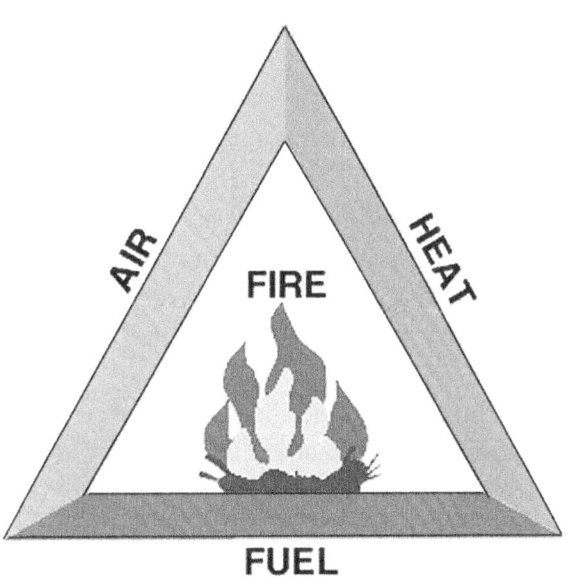

TRAILER INFORMATION

ALWAYS STABILIZE THE TRAILER BEFORE YOU ATTEMPT ANYTHING ELSE. 1,000 POUNDS OF PANICKED ANIMAL WILL CAUSE SOME MAJOR SHIFTING.

Trailers come in a variety of floor plans from the basic one-horse straight load to the combination trailer and living quarters. Knowing a little about trailer construction and layout will aid in deciding your method of extrication. Here are a few of the variations:

Straight load: Horse(s) stands with his nose at the front of the trailer and his rear at the back of the trailer. Most often for two horses, although a 4-horse size is available. Dividers between horses. Usually, the exit for the horse is at the rear, although some like *Brunderup* have forward exits as well. Two horses make a two-layer rescue. If the horses have on saddles or harness they may be tangled or hung up on the divider.

Slant load: May look the same as the Straight Load or it may have several windows along the left side. Horse stands at an angle, usually tied head-forward to the left side. Dividers between horses are usually removable. These trailers can hold 3, 4 or more horses depending on the length, although a single horse could ride without being tied.

If the trailer lands on the side where the horses are tied, this will leave them attached to the trailer, head down, rear in the air. If the trailer goes down on the other side, horses may be hanging from their lead ropes. Be sure to cut these ropes immediately.

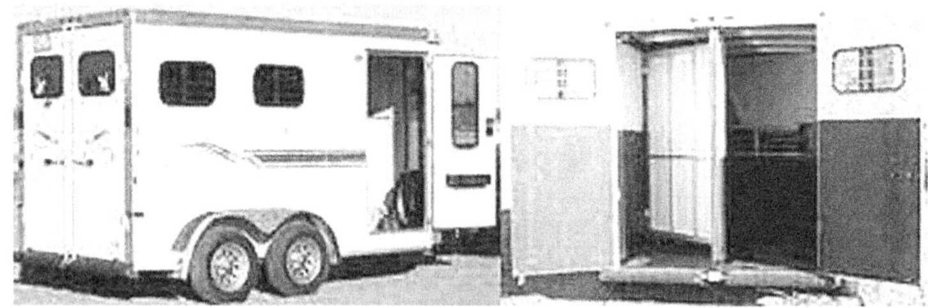

Tandem load: This rarely seen trailer is a straight load for two horses, except the horses ride one behind the other instead of side by side.

Stock trailer: Often has no interior dividers but could have several different configurations of dividers. The solid sides have bars at the top instead of windows. Horses are usually tied on the left side, although a single horse may ride without being tied. The above information for Slant Loads applies. Could be holding livestock or exotics instead.

Stock trailer vs. Slant load trailer

Van: This is a box on a truck chassis. There is usually a ramp in the middle of one side. Often used for 4 horses, can also contain 6 horses. Like the straight load, the horses are in line, except two or three will be facing the other two or three, with the loading area between them.

Horse Van with Ramp

Doors: Can open from the middle, the side, slide open halfway or open to become a ramp. May have a full-sized "escape door" at the front on the right or both sides. Some "escape doors" are through the front of the trailer. Most of these "escape doors" are for the owner to escape, after leading his horse in. You can use them to get to the horse, but be careful as the horse may see the opening as his best escape route, making a hard situation a lot trickier. Some trailers have a ramp in the front of the trailer as well as at the back.

Door configuration can indicate the floor plan of the trailer. CAUTION: Ramps are potentially dangerous because they are spring-loaded and very heavy. Combination back door/ramps can be hard to open, and can be unstable due to their springs.

Dividers: Can be a single pipe from front to back, a full wall, or anything in between. Can be removable or fixed.

Tack Rooms: This is a storage area. Can be located at the front of the trailer, under the feeding platform in straight load trailers, or be full height. May have access from both sides or just one.

May also be located in the rear. If this is the case, you will have a harder time extricating horses. You can pull them out more easily if the trailer lands "tack box side up." If the trailer lands with the tack room down, and if you can't right the trailer or remove the box, the easiest way to get horses out is by making a ramp over the tack room. (These are common in the trailer/RV combinations.)

Windows: Some trailers have small windows at the front for access to the feeding platform. Some trailers have windows all along the sides....slant loads will often have a window for each stall. Stock trailers generally do not have windows and have open, slatted sides. Horse may try to escape from even the smallest window.

Floors: Many trailers have wooden floors – some attached, some not. Some are installed over reinforced welded wire. On top of the wooden floors, many trailers have thick rubber mats. They may also have straw or shavings on top of the mats. Some trailers have mats half way up the walls as well. Some trailers have a solid metal floor.

Materials: Steel, aluminum, fiberglass, wood. Fiberglass can be used on the roof or the entire trailer. Wood is used for floors and occasionally for sides. Aluminum trailers are joined to their axles with steel. Here are some common dimensions:

Top:	20 gauge flat steel; .032 - .50 gauge sheet aluminum
Frame:	14 – 18 gauge steel; ½ inch aluminum tubes
Skin:	16 – 18 gauge steel; .063 gauge sheet aluminum
Bottom:	2x8 inch wood; 7/16 inch extruded aluminum; 2 inch I-beams

Hitches: If you need to call a tow vehicle it's important to know if the trailer is bumper pull, gooseneck or 5th wheel (both have attaching mechanisms in the bed of the truck) or tractor trailer. Couplers and balls will most commonly be either 2 inches or 2 5/16 inches.

DANGERS: Many trailers have batteries at the front to power their braking system. There is also an electrical connector to the towing vehicle. Combination trailers (horses and living quarters) will also have propane cylinders, gasoline generators, and stout electrical connectors.

BONUS: Look for lead ropes, halters, blankets, rescue equipment, first aid box, chocks for wheels and other useful items in the tack room or RV areas.

TYPICAL WEIGHTS of conventional trailers, courtesy of, and copyrighted by, Large Animal Rescue Co.

SIZE	AVERAGE	RANGE
2-horse	3550 lbs.	3000-4200 lbs.
3-horse	4400 lbs.	3300-5000 lbs.
4-horse	5000 lbs.	4000-7000 lbs.

Stock trailers are referred to by length:

12 foot	3200 lbs.	2500-3600 lbs.
14 foot	3400 lbs.	3000-4000 lbs.
16 foot	4050 lbs.	3100-5200 lbs.
18 foot	4800 lbs.	3500-7000 lbs.

TOOLS FOR EXTRICATION
The good, the bad, and the ugly

Some tools will create more havoc than help. Use the tool that causes the least amount of panic in the horse .... the least vibration, noise and light. Trailers are caves, reverberating and amplifying sounds, and power tools will cause them to shake. If you are going to use "bad" tools, tell the vet. She may want to sedate the horse to keep him from going into shock.

When you cut a hole in a trailer to remove a horse, don't just remove enough material to get the horse out. Cut off the whole side or top so there is room for humans and for the horse.

When working with power saws and cutting torches, make sure the horse is protected from sparks and shrapnel by covering him completely with a blanket. Use only natural fiber covers to ward off sparks. Synthetic material could melt into the horse's skin. Also consider this: if you're going to cover the horse's face, make sure there is enough air to keep a 1,000+ lb. animal breathing! Do not block his nose. He is an obligate nose breather. If covering the horse to protect him from sparks, consider using fans in the box of the trailer so the horse doesn't overheat.

Covering the horse's eyes may prevent him from panicking at the sight of sparks.

With the help of the Equine Rescue Techniques' instruction manual, we've listed some of the more common extrication tools.

When you look for hazards, remember that there may be extra feed or bedding in the tack compartment as well as the box of the trailer.

PORTA POWER: This is a good tool and is recommended. It causes the least amount of noise.

HYDRAULIC POWERED RESCUE EQUIPMENT: Good stuff. They are powerful and relatively quiet. Place the power unit as far away from the horse as possible.

HAND TOOLS: Saws, hammers, axes, battering rams, sledgehammers all create noise and vibration. The lesser degree of stress by not using power tools needs to be weighed against the speed the power tools afford.

GENERATOR: Place as far away from the scene as possible to help reduce some of the noise, smell, and vibration.

AIR CHISELS: The noise, vibration and possible sparks can panic the horse. If time is the paramount consideration, use the air chisel, but sparingly. It **WILL** shake the trailer!

CUTTING TORCHES: These are bad. The smell and sparks may cause the horse to panic. Fire is a natural enemy of horses and he may try to flee. Use blankets or otherwise shield the horse if you need to use these.

ROTARY POWER SAW: This is an ugly tool and is not recommended. As the saw cuts through metal it produces a shower of molten metal sparks that can ignite flammable vapors or combustible materials. These same sparks will injure the animal.

HOSE DIMENSIONS

All sizes are in inches. For centimeters, multiply inches by 2.54. Example 1.5 inches = 3.81 centimeters.

SINGLE JACKET						
Size	1.5	2.5				
Outer Dimension	1.63	2.8				
Flat Width	2.56	4.4				
DOUBLE JACKET						
Size	1.5	2	2.5	3	4	5
Outer Dimension	1.88	2.3	2.88	3.3	4.37	5.37
Flat Width	2.95	3.61	4.52	5.18	6.86	8.44

Thanks to Julie Young, Niedner Hose, Coaticook QC, Canada www.niedner.com

BUILDING AN EQUINE AMBULANCE

Roger Lauze of The Massachusetts Society for the Prevention of Cruelty to Animals (MSPCA) at Nevins Farm writes, "We are a unique property incorporating an animal care & adoption center for small animals, an Equine & Farm Animal Center (one of the only open door facilities in the country), a Humane Education Program featuring a very popular Children's Summer Camp, the Hillside Acre Pet Loss Services and our nationally recognized Equine Safety & Ambulance Program.

"Our trailers must do double duty, functioning as regular horse trailers for the other programs at Nevins Farm as well as providing support for our Ambulance Program. This program has four components: the on-call Equine Emergency Response & Rescue service, the Sport Horse Ambulance service, our Equine Safety & Rescue Training service, and distribution service for the Original Rescue Glide.

"The biggest consideration when we designed our trailers was that they had to be horse trailers first and ambulances second. They had to be able to function in New England weather and road conditions in emergency situations. Therefore they had to be light and maneuverable, have front unload capability, and still hold all of our equipment. At the program's inception, we considered all of our options when it came to vehicles. We thought about four horse trailers to give us more room and even used a four horse in the early stages, but found it was too heavy and hard to maneuver off road. Hydraulic floor trailers and trailers with large water tanks were considered but our basic philosophy of light and maneuverable held out and we decided on two horse front unload goosenecks. Our trailers are purchased off the lot, not specially designed for us, and require only slight modifications.

"Why front unloads and goosenecks? In our experiences we found that some horses that are injured and standing find it difficult to back out of a trailer so the front unload makes it easier for them. Although adding the front unload lengthens the trailers, the gooseneck feature allows us to offset the added length by making them more maneuverable. The inside deck of the gooseneck provides a place to house our equipment and makes each trailer a complete unit so no matter which of our trucks we use, all of our equipment will be available. Our trucks are ¾ ton four wheel drive; this allows us the pulling power and stability for our trailers and bad driving conditions.

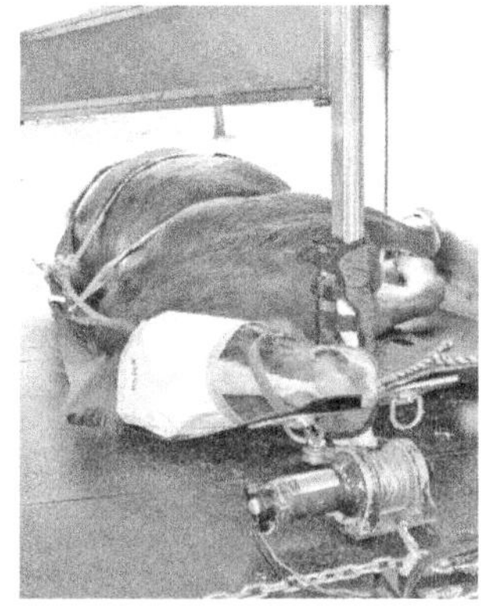

"The Original Rescue Glide is designed to fit in a two horse trailer without removing the center post or divider, for two reasons. First, the trailer is always set up to transport horses and does not require change when we use the Rescue Glide. Second, with the notch in the front of the Rescue Glide fitted around the center post, it places the horse's head on the other side of the breast bars, thereby giving the veterinarians a safety zone to work in. The one condition is that the horse fit completely on the Rescue Glide. Only three times in the sixteen years we have been moving down horses were horses (three very large drafts) longer than the Rescue Glide. In these cases we had to remove our center post and divider (storing them on the gooseneck deck), and then pull those horses into the trailer.

When we purchase trailers we make sure they are designed so the center post and divider can be easily removed. We make two small changes to the trailers. Two eye bolts are inserted through the floor in the front of the trailer with metal plates under the floor so they can't pull through. We then attach a chain between the eye bolts and this gives us an anchor for our winch. The second modification is to attach three wooden 2"x4" planks to the side of the trailer over the gooseneck deck to hold the Rescue Glide and Slip Sheet; this allows us easier access to the Rescue Glides and more space for equipment on the gooseneck deck itself. By housing the Rescue Glides above the gooseneck deck we are able to carry more equipment including the Anderson Sling and the UC Davis Large Animal Lift.

"Our Rescue Glide system makes it possible to make most trailers into equine ambulances without losing the original function of the trailer. Specialized and larger trailers may have a place in equine rescue but in our program, where light weight and maneuverability is a necessity, they wouldn't work for us. A trailer that can get as close as possible to the injured horse in bad driving conditions is the most important factor when we consider trailers."

TRUCK ND TRAILER INVENTORY

GLIDE KIT

Blinder
Ropes
Ratchet Straps

2 Pairs of Hobbles (+ 1 spare set)
3 Locking Carabineers

BAG #2

Antisweat Sheets
Wunderwear

BAG #3

Bell Boots
Davis Boots

BAG # 4

BlinderCotton Hay Net
Halters
Lead Lines
30'Cotton Ropes
Two 18' Nylon Rescue Straps

SPLINT KIT

2 Kimzey Splints
Duct Tape
Vet Wrap
Elastic Tape
Leg Wrap Sheets
Cotton Wraps
Polo Wraps
Splints

ANIMAL FIRST AID KIT

Ear Plugs
60 cc Dosage Syringes
Eye Ointment
Rubbing Alcohol
KY Jelly
Thermometers
Roll On Insect Repellent
Vet Wrap
Saline Solution
Hydrogen Peroxide
Furazone
Latex Gloves
4"x 75" Bandages
4"x 4" Gauze Sponges
Size 4 Diapers
Bandage Scissors
Hemostat
Elastic Tape
5'x9' Dressings

TOOL KIT

Assorted Screwdrivers and Wrenches
Utility Knife
Hammer
Tape Measure
Ratchet Set
Duct Tape
Hoof Nipper
File
Leather Punch
Fence Tool
Wire Cutters
Hack Saw
Assorted Rings, Hooks, Snaps
Tree Saw
Bolt Cutter
Monkey Links

OTHER ITEMS IN TRAILER

Rescue Glide
Marine Battery
2 Winches
Screens
Battery Extension Cords
Come-a-Long
Bucket, Scraper, Sponge and Fly Spray
Human First Aid Kit
Cooler
Water Container
Snatch Blocks
Crow Bar
U C Davis Large Animal Lift
Horse Recovery Hood
Anderson Sling
2 J Hooks

Our thanks to Roger Lauze of MSPCA for writing this section

RESCUE EQUIPMENT

All equipment has different strengths and weaknesses. If you are interested in purchasing rescue equipment, please research what equipment would be best suited for your needs.

Warning: Please do not try to use this equipment if you are not appropriately trained.

These harnesses are designed to be used for a very short amount of air time; maximum ten minutes for the Figure Eight harness and five minutes for the Two Point harness.

USING WHAT YOU HAVE

RESCUE STRAP

A rescue strap can be made from 2-½, 3 or 4 inch fire hose or 3-6 inch double ply webbing. You will need approximately 20-30 feet of it. Cut "eyes" at each end of the fire hose, or make large loops using secure knots in the webbing. **NOTE:** Narrow straps can cause tissue damage. Please use the widest material available, or place straps side by side and wrap together. If possible, pad areas of sensitive skin such as inner legs and shoulders.

VERTICAL LIFT HARNESS

Simple but effective for short-hauls is the figure eight harness made with about 50-75 feet of single jacket wild land hose, 3-4 inch double ply webbing, or soft cotton rope. This harness distributes weight where it usually goes--the four legs. But even with proper weight distribution, severe injury can result from hanging in the harness more than 10 minutes. **Do not use this harness with a helicopter**. Pad delicate skin areas abundantly or the horse will constantly shift to relieve the pain.

HOW TO MAKE A FIGURE EIGHT VERTICAL HARNESS

A temporary harness can be used to support a horse when he is unable to stand on his own, to help him to his feet, or to perform a vertical lift. **Do not use this harness for longer than ten minutes. Very secure.**

1. Mark the half-way point on the strap. Center half-way mark over withers (body end of neck where it joins the back)

2. Bring ends forward and tie a half-hitch knot at the sternum (base of neck, below the "soft spot" or thoracic intake area -- we'll call this part the "neck loop")

3. Run the ends of the rope between his front legs; up behind his front legs

4. Cross the ends over the horse's back, forming an X behind the withers

5. Bring the ends down the flanks, and run inside and between the hind legs. Be careful to **not** cross the ends between the legs. Position carefully!
6. Bring the ends up and tie a half-hitch above the tail

7. Feed the ends under the X and under the "neck loop". Once the harness is on the horse, go back and tighten the straps

8. Bring ends rearward, back OVER the X, tuck between the two straps at the back and bring forward to the "neck loop" again

Wind the ends around the straps several times – each end going in the opposite direction – between the neck and the withers, until a handle is formed. This handle can be used for lifting. Tie the ends off with an overhand knot

Tie a "cutaway strap" between the harness and the lifting device. In the event the lift goes wrong you can cut the strap and leave the harness in place on the horse

9. PLEASE pad under the strap where it will put pressure on delicate tissue, such as shoulders and inner legs

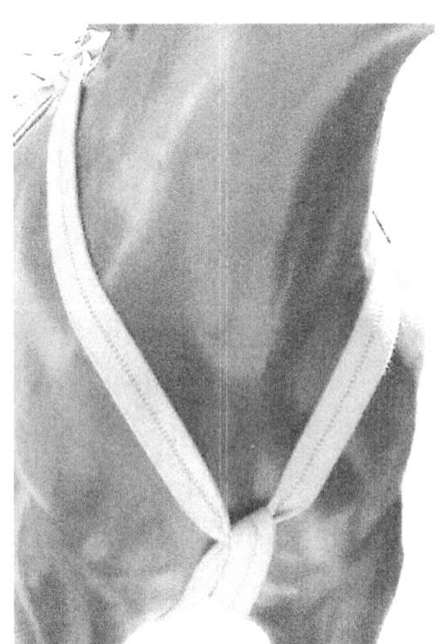
Steps 1 and 2

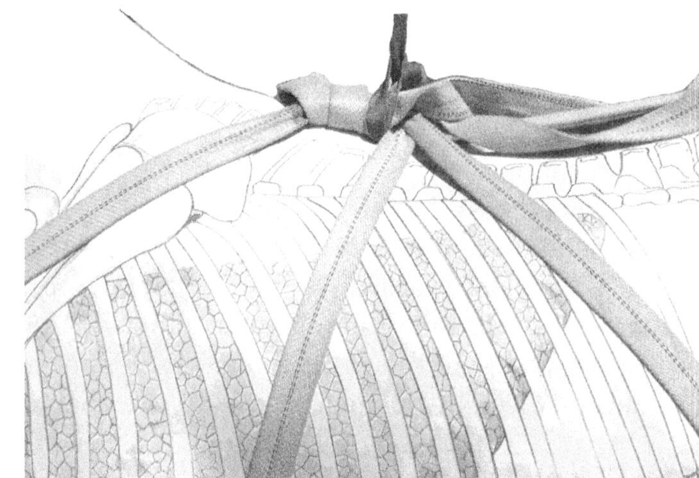
Side view at withers showing cutaway strap

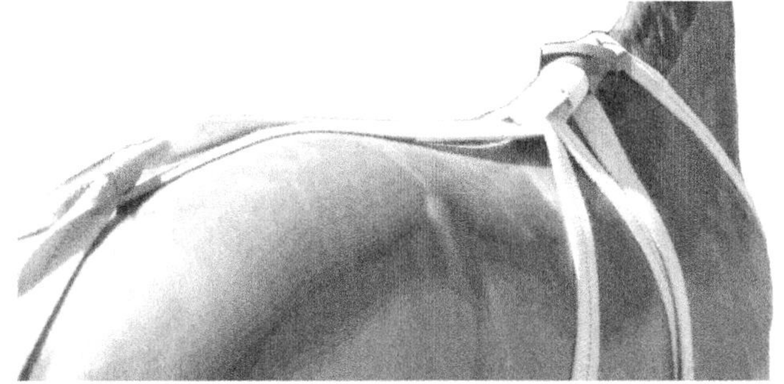
View from the back showing Step 6

HOW TO MAKE A TWO POINT VERTICAL HARNESS

Effective for short-hauls (five minutes), this harness is made of four sections of hose, webbing, two-ply tow straps with flat ends, or a combination. One vertical "body strap" goes behind the front legs and one goes in front of back legs (being aware of "male parts"), then one horizontal strap goes around the horse's chest and one around his rump. Very secure.

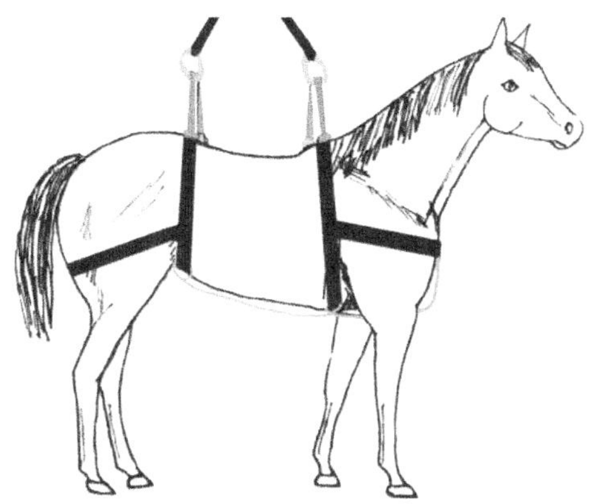

If you are using pieces of hose or webbing, tie knots in the ends of each vertical strap (each knot you tie will take up 3 feet of strap), and then connect them together over the horse's back with about 3 feet of utility rope or webbing, tie the ends to make a loop, and then, making girth hitches at both ends (so they are double), slip these over the knots. This can sit on the horse's back until you're ready to lift.

Once these are in place, tie a strap across the horse's chest and to the front strap. Tie a similar strap across the horse's rump and attach to the rear "body strap" about level with the bottom of his chest.

Next, using webbing, fold it in half and at the bight, create a girth hitch at the center of the chest strap. Pass the ends of the strap between the horse's front legs and wrap the free ends around the front belly strap and tie a water knot. Connect the front and back "body straps" underneath the horse by continuing this webbing to the back strap and tying with another water knot, bringing the two "body straps" approximately two feet apart.

For added safety, take another approximately eight foot length of webbing, fold in half and put a girth hitch in the center of the rump strap. Pass the webbing ends between the rear legs and, separating it so it will avoid the genitalia, tie to the rear "body strap".

When you're ready to lift the horse, connect the vertical "body straps" to the lifting straps in two steps.

1. Connecting straps: Using two 3 foot lengths of rope or webbing, make loops and create girth hitches at each end of each loop. These

can be slid over the knots of the body straps. Pull up in the middle and attach these straps to your lifting strap with carabiners. Make one for front and one for back body straps.

2. Lifting strap: Using an 8 foot length of rope or webbing, make a loop and fold in half, making a double-layer loop. At the bight, tie an overhand knot (web) or a figure eight on a bight (rope) leaving a couple inches of loop at the top. This is then attached to your lifting mechanism with carabiners or Delta Links (triangle screw links). This method prevents tri-axial loading of the carabiner.

From Ranger Magazine, Park Ranger Issue 3: "Improvised Rescue: Large Animal Rescue Using Fire Hose or Cargo Strap" by Tim Collins, Santa Barbara Humane Society (see Commercially Available Equipment section for Tim's Specialized Rescue Equipment)

You can also use two-ply tow straps or tree guards/winch extension straps from a 4WD recovery kit for the two vertical pieces, and webbing or harness breast collars for the horizontal pieces. Some responders in the field prefer to use a spreader bar between the vertical pieces; some contend it is too cumbersome to carry into the field.

EMERGENCY ROPE HALTER

Utility rope can be used to make an emergency rope halter or to tie onto an existing halter. Emergency halters can be severe and damaging if used improperly. Please be very gentle when using one and NEVER use to tie the horse to an object. To make a halter, follow these steps:

Using a soft, long rope make a loop at one end. Put the loop end over the horse's head, up behind his ears (at the poll). The loop should hang down far enough that it will end up centered under the horse's face.

Taking the standing section of the rope, push it through the loop to form another loop. Put this loop over the horse's nose and GENTLY tighten and adjust so that the first loop is centered under the face and the second loop is below the cheekbones.

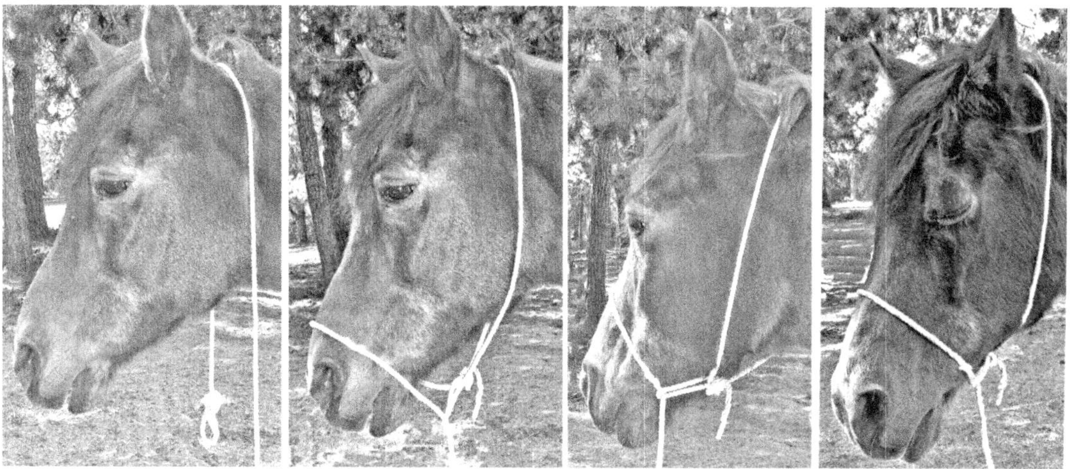

About our model: There are several pictures of Chief in this book. He crossed the "Rainbow Bridge" about 6 years ago, but he lived until a month shy of his 34th birthday.

Chief was a rescue who came to live with me when I first became involved with rehoming horses. He was an Egyptian Arab from the Hearst family breeding program. He stood 14.2hands and his conformation was perfect. Unfortunately, he had a stifle injury and his owner no longer wanted him.

Chief was dutiful toward his work with handicapped children, enjoyed teaching non-horse people who were afraid, and had a wicked sense of humor. He had two particular friends whom he loved very much. One was a five year old girl (Amber) who would drag a halter and lead into the pasture where Chief would put his nose on the ground so his friend could halter him. She would take him to the round pen where she would ride for awhile, then slide off and sit on his feet, playing in the sand. Chief would doze off and the two would be like that for hours. The other friend was my mom who had been terrified of horses until they became a big part of my life. When mom was out in the yard and the horses were loose, Chief would seek her out and hang his head over her shoulder. He was the first horse she rode by herself (not being led) and she was quite proud of that.

Chief was not a fan of his own species. He'd chase the other horses. One day a trainer/friend (Amber's dad, Dave) was riding a young colt (Austin) toward the barn and Chief thought he'd cause some mayhem. He ran at Austin with his teeth bared, mouth wide open. Austin jumped and Chief's teeth enveloped Dave's leg. Chief didn't bite down, of course, and he was horrified. His eyes got really big and he ran into the corner of the field and stood, facing the corner, with his head down for the better part of an hour.

I miss this boy. He was a wonderful friend and mentor.

COMMERCIALLY AVAILABLE EQUIPMENT

ANDERSON RESCUE EQUIPMENT

Care for Disabled Animals/CDA Products
PO Box 53 Potter Valley, CA 95469
(707) 743-1300
fax (707) 743-2530 www.andersonsling.com

Vertical Lift Harness
The Cadillac of rescue equipment is the Anderson Sling, designed by Charles Anderson. He was assisted by Dr. John Madigan, who gave the sling hours of practice at the veterinarian school hospital at the University of California, Davis. It is used for animals who are too weak to stand, those with injured legs, and for animals coming out of anesthesia, even for long periods of time. This harness **can** be used with a helicopter. The fastening straps require training and practice.

Rescue Glide
The Rescue Glide is a molded sheet of 8' by 4' high-density plastic. The front end is turned up to facilitate movement over uneven ground. Built in handles along the sides can be used with straps to contain a horse on glide. Comes with two tough, slick plastic slip sheets to help it flow over the stickiest surface.

Also available: equine lifting hobbles to fit onto the Rescuer Frame of the Anderson Sling (not to be used with a helicopter), small animal walkers to help keep foals on their feet, full size fiberglass horse for training, Dura-Pic Tripods, and hydraulic frames used in rehabilitation settings.

PHIL CANDY

Blacksmith Tel. 02380 616056 Mobile 07891 826195
Bishopstoke, Hampshire, UK

Strop Guide
Designed by Hampshire Fire and Rescue Services (UK), the Strop Guide enables the placement of straps and harnesses under a recumbent large animal in a safe manner, providing additional reach and minimizing further injury to the animal. Made of sprung steel, it bends around an animal, enabling a strap to be attached at close quarters. Dimensions: Length 2160mm (7'). Width: 40mm (1.5"). Steel thickness: 4mm (.16") with a 'D' Shaped handle at one end and a hook and steel ball (providing attachment for the strap) at the other end.

COLLINS RESCUE EQUIPMENT

Timothy Collins with the Santa Barbara Humane Society
Santa Barbara, CA
(805) 687-1328 timcollins@sbhumanesociety.org

Tim was one of the pioneers of Large Animal Rescue in the USA and he has created and adapted many items of rescue equipment for horses.

Collins Rescue Mat
This 8' x 8' reversible mat has tie-down loops and is made of webbing-reinforced trampoline mat material. The tow bar separates, allowing you to pull a horse through a narrow spot. By reversing the bar placement on the mat you can reverse the direction of the pull. The Collins Rescue Mat can be used for a horizontal drag or on a steep hill. It rolls up to fit in a duffle bag and can be carried by one person.

Collins Back Country Rescue Mat
The Collins Back Country Rescue Mat is 3'6" (1m) x 7' (2m), has webbing tie-down loops and is not reversible. The steel bar separates and the whole thing fits into a backpack.

Collins LAER Harness

Weighing less than 14 kilos and fits in a backpack. The loops sewn along the 112 mm-wide (4") tubular webbing allow rescuers to secure animals of various sizes. Connect the pieces using 25 mm (1") webbing.

Collins Rescue Anklets
Two padded, adjustable loops made of 5 cm webbing and attached to a D ring. The set of anklets comes complete with a steel lift bar and is used for rollovers, for lifting a large animal feet first allowing for quick application and removal, and can be used with a rescue mat to restrain the animal. The anklets are not used in conjunction with a sling. **Not for use with a helicopter.**

Collins Rescue Strap
The Collins Rescue Strap is used for a horizontal forward pull or backward pull, and to assist a standing horse when on steep terrain when you can stop him and hold him in position while he rests. The Collins Rescue Strap has diagonal slots sewn in to prevent wrinkles putting pressure on the horse and it has a Velcro keeper to prevent the loose end from flopping around.

Water Rescue Balloon
The Water Rescue Balloon is attached to a rubber or flexible hose through which air is pumped. Use a strop guide to push the balloon down the side of the animal until it's beyond the midway point. Pump air to inflate the balloon, bringing the line to the surface and allowing you to attach a larger rope or strap to secure the animal for your rescue. If you need to place several straps under the animal, connect a couple of lines to the loop to save time. Add the air slowly so as to not explode the balloon or spook the horse.

Mud Tubes
The Collins Mud Tubes are 4½ - 5' lengths of rigid plastic piping that have tapered bottoms to allow air to escape even when the ends are against a hard surface. The Tubes have valve stems at the top to allow air or water to be injected and prevent the loss of air when the air tank is removed. Injecting air and water at the same time creates slurry that lessens the suction of mud.

HÄST PUBLIC SERVICE CORPORATION

Kathleen Becker, DVM, MEng, President
2787 Floyd Highway South　　Floyd VA 24091
(804) 286-0832　　http://rescue.hastpsc.com

Becker Sling
The Becker Sling is sold as either a total kit or individual replacement parts. The ultimate load on the aluminum spread bar is just under 10,000 kilos at failure and the Becker Sling should easily handle large animals up to 900 kilos with complete safety. **This sling is not to be used with a helicopter.**

Rescue Frame
This frame is simple and very strong, and can be utilized where a rescue is required but one either does not have access to a crane or the area is inaccessible to heavy equipment. Constructed entirely of aluminum alloy. The usable height of this frame, from the ground up to the ring, is 18 feet 6 inches (5.67m), This "family" of frames includes a bipod, tripod, and monopod.

LARGE ANIMAL RESCUE GLIDE EQUIPMENT (L.A.R.G.E.)

Ben McCracken　　(864) 270-1344
The Upstate of South Carolina　　www.rescueglides.com
Benmccracken@rescueglides.com

Manufacturer of high quality custom rescue glides since 2002. From the smallest to the largest animals, they will custom fit your glide to your specific needs. In over ten years, not one has failed.

LIFTEX INC.

48 D Vincent Circle　　Ivyland, PA 18974
(800) 478-4651　(215) 957-0810 or fax: (215) 957 9180
www.liftex.com　　info@liftex.com

Vertical Lift Harness
The Liftex Vertical Lift Harness provides full abdominal support and secures the animal from escape at front, back and top. The spreader bar is incorporated into the sling and does not need to be supported by rescuers. Comes with hobbles. **This harness cannot be used with a helicopter.**

MASSACHUSETTS SPCA

Roger Lauze, Equine Rescue and Training Coordinator
Nevins Farm (978) 687-7453 Ext. 6124
400 Broadway Methuen, MA 01844-2052
www.mspca.org lauze@mspca.org

Original Rescue Glide
This is the original Rescue Glide designed, manufactured and used by professionals for over fourteen years and has been an integral part of the MSPCA at Nevins Farm Equine Ambulance programme since its design. MSPCA has moved downed horses off paddocks, out of bogs and through stable doorways into its equine ambulances using the Original Rescue Glide. The most flexible glide available, it will never lose its original shape. All metal pieces are aluminium to prevent rusting and screws are countersunk for safety.

MFC SURVIVAL LTD

Naval Yard www.mfc-survival.com
Tonypandy +44(0) 1443 433075
Rhondda Cynon Taff CF40 1JS UK
Partners with **Fire Rescue Safety Australia (FRSA)**
www.frsa.com.au National Call: 1300 4 RESCUE
enquire@frsa.com.au

Animal Rescue Path
Developed to provide a compact and lightweight solution for animals trapped in mud. The inflatable Path creates a stable, cost-effective pathway. It is easily inflated, low pressure, lightweight, compact and portable. Comes in 5 meter (16.4') and 10 meter (32.8') lengths.

Mud Lance
Durable and versatile stainless steel lance used to inject either water or air into mud, allowing the animal to be rescued quickly and safely. Comes with .5 meter (1.6') and 1 meter (3.3') lengths.

Water Rescue Stick
For use during mud rescues. Extendable from 190cm (6.2') to 467cm (15.3'), it has a large hook on one end and is secured with a clamp lock.

RP3 Rescue Platform
Ideal for shallow water or mud rescues. Where necessary, it can be operated behind a jet ski or fast rescue craft and is easily paddled or line transferred, carrying a variety of rigid stretchers. Lightweight and portable, rapid inflation and deployment, very stable, packs into a light compact carrying case, easily inflated.

RESCUE CRITTERS

Thales & Co.LLC (818) 780-7860 Fax: (818) 780-1078
15635 Saticoy St. Unit A info@rescuecritters.com
Van Nuys, CA 91406 www.rescuecritters.com

Horse Mannequin
'Lucky' is a life-sized horse mannequin with fully articulated limbs; he is a realistic training weight and stands at 15 hands. He will accept standard horse harnesses, rescue glides and gear, can be used in all weather, mud and water. Lucky comes with all the accessories needed to assemble him, and a carry bag for his connecting rods, knobs and washers. All his components are engineered to be rust-proof. His price does not include the reusable shipping crate that is also used to store and transport him.

RESQUIP LIMITED

Unit 1, Castle Court: +44 (0) 1938 558888
Leighton, Welshpool www.resquip.com
Powys, SY21 8HH WALES

Horse Mannequin
The mannequin is fully articulated, stands 15.2 hands, and is the only one that stands up. Comes in any color but white is preferred as it minimizes heat build-up if the horse is left out in the sun, and clearly shows the precise position of slings etc. Asia Pacific distributor is MaryAnne Leighton www.equineER.com maryanne-leighton@bigpond.com US distributor is Dr. Rebecca Gimenez www.tlaer.org/Rescue_Randy_Home.html

Rescue Glide System
Made of high-density polyethylene, easy to use, store, transport, and assemble on-site with two quick connectors. Mats have matching holes on all four sides for attaching and pulling the galvanised steel glide bar – with a kick-up for riding over rough terrain – on any side. If an animal is too big for a single mat, join two mats together. The straps make good use of the animal's anatomy by distributing the load around the shoulder and hip (not over them), thus avoiding local pressure on the muscles while keeping the patient steady and preventing it sliding off the glide. Asia Pacific distributor is MaryAnne Leighton at 0429 119 084 or maryanne-leighton@bigpond.com North American distributor is Michelle Staples at michelle@redjeansink.com

Eburn Vertical Lift Sling
Designed by experienced emergency responders, the simple sleek design allows for quick application and release. The three anchor points on the spreader bar (instead of the usual two), using simple, quick connections, gives extra versatility. Used for two-point vertical lifts and single-point hobbled lifts. Competitively priced. **NOT FOR USE WITH HELICOPTERS.** MaryAnne Leighton is the Asia-Pacific distributor. Asia Pacific distributor is MaryAnne Leighton at 0429 119 084, or maryanne-leighton@bigpond.com . North American distributor is Michelle Staples at michelle@redjeansink.com

SHANK'S VETERINARY EQUIPMENT INC

505 E. Old Mill Street www.shanksvet.com
Milledgeville, IL 61051 USA Jennifer@shanksvet.com
(815) 225-7700 or fax: (815) 225-5130

Padded Hood
This adjustable foam hood comes in three sizes, and has ear and eye holes and removable blindfolds. If you already have a hood, blindfold kits are available. The padded hood is designed to fit over a halter to protect the top and sides of the head.

TLAER, INC

TLAER Int'l Training Facility
1787 GA Hwy 18 East www.tlaer.org/store.html
Macon, GA 31217 (214) 679-3629

Nickopolous Needle
100% stainless steel; open circumference (fits medium or draft horse); ring attachments on both ends; connectors for hose attachment using either air or water. Crescent shaped, it encircles the animal's torso. A pilot line can then be attached to the tip to be pulled under the animal's torso.

Hampshire Strop Guide
Cold rolled steel; 6' long x 3" wide with handle at one end, the strop guide bends around a recumbent animal, enabling a strap to be attached at close quarters. A line attached to the tip is then pulled under the animal's torso to encircle the body.

Mud Lances
100% stainless steel (1/2 inch); each set consists of 4 mud lances; connectors provided or use with water for injection at idle pressure (60-80 psi) or air injection at 60 psi (customer choice). Used to break the suction effect of the mud.

TRAINING RESOURCES

Large Animal Rescue graduates and friends

WITHIN NORTH AMERICA

CODE 3 ASSOCIATES, INC.

Krista Kurvers, Training Director
1530 Skyway Dr.
Longmont, CO 80504
(303) 772-7724 fax (303) 485-6210
www.code3associates.org
kkurvers@code3associates.org

Code 3 Associates offers training in Technical Animal Rescue including water rescues, Equine Investigations, and Biosecurity and Zoonoses; plus Equine Investigations Academy. Will dispatch their Mobile Command Center to disasters throughout the country.

TIMOTHY COLLINS

Santa Barbara, CA
(805) 687-1328
timcollins@sbhumanesociety.org

Technical Horse Rescue Specialist. Hands-on, eight hour classes.

- Starting an Equine Emergency Rescue Group, which covers personal emergencies, evacuation, trailer work, and community assessments among other things
- First Responder Emergency Rescue Training, which covers disaster management, emergency handling of horses, specialized equipment, and different types of rescues
- Self Preparedness for Horse Owners, which covers emergency handling of horses, types of individual emergencies, facility assessment, how to plan, and trailer work

MICHAEL CONNELL

Reno, NV
(775) 857-2769
mconn009@gmail.com

Technical Large Animal Rescue Instructor (Nevada Division of Emergency Management); Technical Large Animal Rescue Specialist. Michael Connell is a certified technician in TLAER, instructing for the State of Nevada Division of Emergency Management. Offers hands-on, sixteen hour classes, veterinary CEU.

- Classes cover the Incident Command System, on scene safety, equipment
- Animal behavior, barn safety, trailer safety
- Veterinary considerations, euthanasia, triage

Da VINCI EQUINE EMERGENCY TRANSPORT LLC

Nicole Ehrentraut, Team Leader/President
PO Box 882, Poolesville, MD 20837
(301) 335-2340 available 24/7 DaVinciEquine@gmail.com

- Large Animal Technical Rescue Squad/Ambulance
- Basic 1- Day Large Animal Technical Rescue Training
- Emergency Preparedness, Trailering Safely and Trail Riding Incident Safety for Horse Owners

DAYS END FARM HORSE RESCUE, INC.

DeEtte Gorrie, Equine Programs Director
1372 Woodbine Road www.defhr.org
Woodbine, MD 21797 programs@defhr.org
(301) 854-5037, (410) 442-1564

DeEtte is certified in Technical Large Animal Emergency Rescue, and a Licensed Equine Humane Investigator. Large Animal Rescue Training is a one-day, hands-on program on how to safely respond to an emergency equine situation. Participants gain an understanding of basic rescue equipment and how to use these tools in an emergency situation. LART Training Specialists, Level II Equine Investigators Academy, Worldwide training

EASTERN KENTUCKY UNIVERSITY

Larry Collins
Eastern Kentucky UniversitySafety, Security & Emergency Management
250 Stratton Building (859) 622-1009
521 Lancaster Avenue (859) 622-6548 (fax)
Richmond, KY 40475-3102 larry.collins@eku.edu

EKU offers two classes each year in the spring semester. The first class is for local responders and students in the fire and safety bachelor's program at EKU. The other class is open to the public and attendees include veterinarians, vet-techs, mounted police and horse owners. See TLAER for description of program.

EMERGENCY EQUINE RESPONSE UNIT

Eric Thompson, Operations Mgr. (913) 522-3064
10516 Reeder St. www.emergencyequineresponseunit.org
DeSoto, KS 66214 mwr604@yahoo.com

The Basic Equine Awareness and Rescue (B.E.A.R.) course is an 8-hour awareness level course that provides basic horse awareness and emergency response for equine accidents. The B.E.A.R. course includes the use of lift mechanisms and an equine sling, and information on trailer accidents, horse barn fires, and horses stuck in mud. Large Animal Rescue Operations (LARO) is an operations level course that includes practicing rescue scenarios, lift mechanisms, and what to do if a large animal gets stuck in mud, on ice, and other incidents. Large Animal Ice Rescue (LAIR) is an operations level course involving ice and water training with an emphasis on advanced horse rescue techniques involving hands-on and classroom instruction.

EERU also serves as national disaster responders with the Code 3 Associates Essential Animals Team (EAST). EERU serves as a large animal rescue team for EAST, and is trained to do small animal rescue disasters. Equine ambulance service in the Kansas City area, hauling to both Missouri University and Kansas State.

LARGE ANIMAL RESCUE CO., INC.

Deb and John Fox
Hollister, CA 95023 www.largeanimalrescue.com
(831) 635-9021 tlar@got.net

LAR curriculum developers and training specialists.

Large Animal Rescue – Operational, a Fire Service Training and Education Program (FSTEP) class approved by California State Fire Training. Also offering an eight-hour course that teaches rescue concepts, scene management, operations and equipment.

IRV LICHTENSTEIN

8480 Limekiln Pike, #914 (215) 576-1641
Wyncote, PA 19095 ilichten1@verizon.net

State and Federal certified, Irv is the LAR expert for 9 counties in PA, His training program includes 22 courses ranging from ICS and search management to vehicle rescue. Responders can learn how to use equipment they carry on their engines, scene management, and mutual aid resources for both LAR and SAR.

MASSACHUSETTS SPCA AT NEVINS FARM

Roger Lauze (978) 687-7453 ext. 6124
400 Broadway www.mspca.org
Methuen, MA 01844 rlauze@mspca.org

Equine ambulance and rescue training specialist. Teaching LAR classes including the use of the Rescue Glide, the UC Davis LAL, and the Anderson Sling. Emphasis on Emergency Transportation and turning a trailer into an ambulance. Distributor of the original Rescue Glide. Oldest equine ambulance service in the country. Ambulance services for national and local sport horse events. Twenty-four hour ambulance service for veterinarians in New England.

JUSTIN and TORI McLEOD - 4HFES and NCSMART, LLC

4Hooves Farm Equine Services and North Carolina Specialized Mobile Animal Rescue Team

Spring Lake, NC www.4hoovesSMART.com
(910) 494-8210 (919) 201-6789

Large Animal Technical Rescue Emergency Response; Equine Event Standby Emergency Response Unit; Equine Transport (Emergency/Scheduled); Deceased Services for Equines and other Livestock
Large Animal Technical Rescue and Emergency Prevention and Preparedness Training - For Emergency Responders, Veterinarians, Animal Control, and Owners/Professionals

RED JEANS INK

PO Box 881 www.redjeansink.com
Buffalo, NY 14201 info@redjeansink.com

RJI offers this book, plus two "Teach It Yourself" classes which are designed particularly for areas that cannot otherwise get training. Horse Awareness and Safety introduces people to horses in both classroom and arena, giving you basic tools to help respond to an incident involving horses. Introduction to LAR for Horse Owners is designed for owners who want to be prepared to remove – or help remove – their horse from a life-threatening incident and gives basic information to help you work with the emergency response system.

RESCUE 3 INTERNATIONAL

P.O. Box 1050 (800) 457-3728 (US & Canada)
Wilton, California 95693 | +001.916.687.6556
www.rescue3.com info@rescue3.com

Water and Rope Rescue Training
This hands-on, 3-day course focuses on rescuer safety, equipment use, and animal safety in rescues, and includes animal first-aid, communication during emergencies, and how to use ropes to maneuver a boat safely to the victim. Upon completion, students will receive Rescue 3 Certification in Technical Animal Rescue.

VICKI SCHMIDT

Maine State Fire Instructor, Frandford Mutual Aid Fire Training Association

955 Buckfield Rd,
Hebron ME 04238
(207) 890-4590
www.frandford.org
troika@megalink.net

Offering Large Animal Emergency Rescue (LAER) training and seminars for owners and First Responders across the United States and parts of Canada. Equine and farm-risk reduction information and activities a specialty. Courses include lecture, hands-on, and reference materials. Custom LAER training designed for your group needs easily arranged. State approved program includes Certificate of LAER Training.

TECHNICAL LARGE ANIMAL EMERGENCY RESCUE (TLAER)

Dr Rebecca Gimenez
1787 GA Hwy 18 East
Macon, GA 31217
(214) 679-3629
www.tlaer.org
delphiacres@hotmail.com

LAR curriculum developer and training specialist
TLAER offers three levels of training – Awareness, Operations, and Technician.

The 2-day Awareness Level course is intended for everyone, no previous experience necessary, and features lectures, PowerPoint visuals and student interaction with minimal hands-on. Each day of the 3-day Operational Level course consists of lecture and hands-on laboratory techniques, plus a 2-hour night operation to practice search and rescue techniques and learn how to use a Rescue Glide for recumbent animals. The Operational Level TLAER course is intended for those who desire a more specialized, technical level of training.

Hands-on training is conducted with live trained animals including horses and a llama and instruction covers the use of sedatives and tranquilizers, chemical restraint, rescue ropes and knots, rescue from stable fires, mud, helicopter and water rescue, and night-time search and rescue. Training covers natural disasters as well as highway mishaps such as overturned horse trailers/floats. Techniques can be applied to all large animals.

WASHINGTON STATE ANIMAL RESPONSE TEAM (WASART)

PO Box 21
Enumclaw, WA 98022
Emergency Number (425) 681-5498
info@washingtonsart.org

WASART responds to disasters and emergency situations involving livestock and companion animals. Technical Animal Rescue (TAR) Awareness training introduces the concepts and techniques of rescuing large and companion animals in distress. Hands on practice with both live animals and mannequins. Use of readily available equipment such as rope, webbing and tarps to help facilitate rescues will be emphasized in addition to demonstrations with specialized rescue equipment.

WAVE TREK RESCUE

PO Box 225
Index, WA 98256
(360) 793-2325
www.wavetrekrescue.com
admin@wavetrekrescue.com

The Technical animal rescue (TAR) watercourse is the only course in the world that brings floods and rivers into the equation. NFPA Compliant Rescue 3 Certified. This course incorporates water and flood rescue techniques with animal behavior to better understand and perform successful animal rescues. It's an intensive three-day, 24-hour class with one day of classroom instruction followed by two days developing and practicing animal rescue skills.

Large animal advanced handling. This course will prepare you for technical rescue of large animals in advanced situations such as: trailer accidents, over embankments, caught in bogs/ mud, and vertical and low angle lifts

JENNIFER WOODS

RR #1
Blackie, Alberta
T0L 0J0 Canada
(403) 684-3008
www.livestockhandling.net
livestockhandling@mac.com

Consultations and training across North America, Europe and Australia on accident response. Since 1998 Jennifer has trained over 10,000 people including emergency responders, truckers, producers, animal enforcement, disaster services, animal welfare organizations, brand inspectors and veterinarians. Classes include: Livestock Emergency Response Training program, Equine Emergency Response, Livestock Behavior and Handling, Livestock Handling for Youth, Disaster Planning for Livestock, Horse Hauling Course, and Emergency Euthanasia. Jennifer is also a certified trainer for Canadian Livestock Transport Program.

OUTSIDE NORTH AMERICA

Australia: QUEENSLAND HORSE COUNCIL

P O Box 1329
Oxenford, Queensland
4210 Australia

www.qldhorsecouncil.com
LargeAnimalRescue@qldhorsecouncil.com
0429 119 084

Queensland Horse Council gives Large Animal Rescue presentations and demonstrations to clubs and groups and conducts one-day information workshops throughout Australia through their Queensland Horse Council Large Animal Rescue Roadshow.

United Kingdom: HAMPSHIRE FIRE AND RESCUE SERVICE

Jim Green, Animal Rescue Manager
Contact through British Animal Rescue and Trauma Care Association (BARTA)
www.bartacic.org

- Fire and Rescue Courses: AR1- Basic awareness of incidents involving animals; _AR2- Animal Rescue Responder; AR3- Team Leader/Instructor; AR4- Tactical Advisor
- Veterinary Courses: VAR1- Basic awareness of incidents involving animals; VAR2- Animal Rescue First Responder; VAR3- Rescue Trained; VAR4- Trauma specialist
- Also provided; Fire Safety on equine establishments- Bespoke course for Riding Establishments Act Veterinary Inspectors; BHS Modules- Dealing with the unforeseen event and Fire Safety on equine and agricultural establishments; Bespoke AR1 awareness for agencies such as Highways Agency, RSPCA

Please contact Jim for training outside the UK as well. Training locations have included: Australia, New Zealand, Norway, Austria, Belgium, and Turkey; with LAR support to France, Sweden and Czech Republic

OTHER HELP and EQUIPMENT

EMERGENCY ANIMAL RESCUE

P.O. Box 2462
Ramona, California 92065
(760) 789-5775

www.emergencyanimalrescue.org
ear@rescueteam.com
(760) 594-0751 24 hr. Emergency number

Emergency Animal Rescue (EAR) does the actual physical rescue of both domestic animals and wildlife. From untangling a hawk wrapped in fishing line in a tree to crawling through storm drains for the baby ducks. In the past twenty + years we have assisted hundreds of people with their pets, the vets that have horses down, the dogs that have gone over cliffs and of course our "wildfires." Contact us either by e mail, Facebook or phone

HUMANE EQUINE AID and RAPID TRANSPORT (HEART)

179 Acorn Hill Drive
Madison, VA 22727
(561) 685-3275 (703)304-5248
(561) 301-1642 (ambulance)

Robin Sweely, Executive Director
www.equineambulance.com
equineambulance@hotmail.com

Non-profit that operates an equine ambulance at major horse shows from FL to NY. Provides humane aid and rapid transport for horses at major horse events.

JOE KEY

www.onlinepoliceexperts.com (540) 377-9966

Consultant/Expert Witness Services in Firearms, Ballistics, Police use of Force.

MARYANNE LEIGHTON

Queensland Horse Council Large Animal Rescue Education
Author of *Equine Emergency Rescue – a guide to Large Animal Rescue*
Asia Pacific distributor of Resquip rescue training equipment
P O Box 1329 www.equineER.com
Oxenford, Q 4210, Australia 07 5573 3974 / 0429 119 084

NIEDNER FIRE HOSE

675 rue Merrill Street (800) 567.2703 / (819) 849.2751
Coaticook, QC CANADA sales@niedner.com
J1A 2S2

SANTA BARBARA EQUINE ASSISTANCE AND EVACUATION TEAM INC.

PO Box 60535
Santa Barbara, CA 93160 www.sbequineevac.com
(805) 892-4484 sbequineevac@gmail.com

A non-profit organization that assists all Santa Barbara County emergency responding agencies and large animal owners in the evacuation, temporary care and sheltering of large animals in time of disasters or accidents. Volunteers are trained and registered Disaster Service Workers. SB Equine Evac provides education and demonstrations for many local public events, agencies and organizations.

SOUTHEAST PENNSYLVANIA SEARCH AND RESCUE

8480 Limekiln Pike, #914 www.sepasar.net
Wyncote, PA 19095 ilichten1@verizon.net
(215) 576-1641

SPSAR has a nine county range, performing searches for both humans and horses, and does the actual physical rescue of both domestic animals and wildlife. Will transport large animals in emergencies. Available for disaster planning and other public education.

USRIDER EQUESTRIAN MOTOR PLAN

moreinfo@usrider.org www.usrider.org
(800) 844-1409

Nationwide member-based organization providing roadside trailering assistance, including towing and roadside repairs for tow vehicles and trailers with horses, emergency stabling, veterinary referrals and more.

WEBSITES

FIRE SAFETY IN BARNS

Author and barn fire safety expert, Laurie Loveman, provides the basic facts on fire safety in horse barns and other livestock housing facilities, and keeps you up-to-date on fire prevention products and practices. Included are Laurie's articles on the subject and a Loss of Animals by Fires Chart. **www.firesafetyinbarns.com**

SAVE YOUR HORSE!

An interactive website, containing information about large animal rescue, stories, instructors, and articles. You are welcome to use the safety articles for your own newsletters and magazines. **www.saveyourhorse.com**

MEMORANDUM OF UNDERSTANDING FOR VETERINARIANS

This MEMORANDUM OF UNDERSTANDING (MOU), between *AGENCY* and *VET*, specifically defines the role *VET* may play during a large animal emergency. This MOU must be officially in place prior to an incident.

Whereas, the parties hereto desire to coordinate a program of large animal rescue by means of this Memorandum of Understanding; and
Whereas, *AGENCY* will act as a primary resource and *VET* will act as a supplemental resource; and
Whereas, it may be necessary that the resources and facilities of mutual aid response agencies be made available to combat the effects of large animal emergencies which may result from, but are not limited to, such calamities as flood, fire, earthquake, and other natural and man-made incidences; and
Whereas, this Memorandum of Understanding recognizes the potential need for livestock and pet intervention and developing animal safety and relocation procedures to be used in future emergencies.

NOW, THEREFORE, it is mutually agreed and understood as follows:

1. The scope and magnitude of the *AGENCY* response will be based on availability of personnel and resources.
2. *AGENCY* may be asked to call for mutual aid from other agencies with personnel trained in Large Animal Rescue.
3. *VET* has been trained in Large Animal medicine and rescue and has knowledge of safe procedures in carrying out such a rescue.
4. In the event of an incident involving the rescue of one or more large animals, *VET* may be called upon for technical advice. If *VET*'s assistance is requested on-scene by *AGENCY*, (s)he will be covered under any liability and disability insurance carried by *AGENCY*.
5. *VET* will not be held responsible for the injury or death of participating *AGENCY* personnel, animals involved, or bystanders; and will not be held responsible for damage to equipment or property involved in said rescue operation, unless such injury, death or damage is proven to be caused by willful neglect or gross negligence.
6. *VET* will be issued identification by *AGENCY* to be used in gaining access to the scene of a Large Animal Rescue incident.
7. *AGENCY* will not be charged for services rendered by *VET*. Similarly, *VET* will not be charged by *AGENCY*.

Signed ______________________________ Date: ________________________
AGENCY Place: ________________________

Signed ______________________________ Date: ________________________
VET Place: ________________________

TO EMERGENCY RESPONDERS

Print Clearly:

I/We, ____, own the animal(s) in this trailer.

Address: ____
____

Phone: Home (________) ___________-______________
Cell (________) ___________-______________

Emergency contact who has legal authority to make decisions on treatment for the animal(s):
Name: ____
Address: ____
____

Phone: Home (________) ___________-______________
Cell (________) ___________-______________
E-mail: ____

Home veterinarian(s):
Name: ____
Phone: Office (________) ___________-______________
Cell (________) ___________-______________
Pager (________) ___________-______________
E-mail: ____

Insurance Co.:
Contact: ____
Phone: (________) ___________-______________

In the event that I/we are incapable of making decisions regarding the health and well-being of the animal(s) in an accident or emergency, we hereby authorize and shall hold harmless a veterinarian to determine the health status of the animal(s), provide emergency health care, or administer an euthanizing agent if the veterinarian determines that an animal cannot be saved.

Signed and Dated

______________________________________ ________/________/________

______________________________________ ________/________/________

Witness: and Date

______________________________________ ________/________/________

This form developed in cooperation with Drs. Tomas and Rebecca Gimenez.

LIMITED POWER OF ATTORNEY FOR ANIMAL HEALTH CARE

[The purpose of this document is to give the person you designate (Your "Agent") broad powers to make health care decisions for your animal(s), including power to require, consent to or withdraw any type of care or medical treatment for any medical condition and to admit or discharge your animal(s) from any hospital, clinic or other institution. This document does not impose a duty on your agent to exercise granted powers; but when a power is exercised, your agent will have to use due care to act for your benefit and in accordance with this form. A court can take away the powers of your agent if it finds the agent is not acting properly. You may name co-agents and successor agents under this form, but you may not name a health care provider who may be directly or indirectly involved in rendering health care to your animal(s) under this power. Unless you expressly limit the duration of this power in the manner provided below, until you revoke this power or a court acting on your behalf terminates it, your agent may exercise the powers given herein throughout your lifetime, even after you become disabled, incapacitated or incompetent. It is recommended that you keep a copy of this document in your tow vehicle, along with a copy of the "To First Responders" document, which provides additional contact information and details on the care and treatment of the animals. If there is anything about this form that you do not understand, you should consult a lawyer.]

LIMITED POWER OF ATTORNEY made this _____day of ________________,
20____.
I/We, (name) ____,
of (address) ____,
(city) ______________________________, (state) _______, (zip code) ____________,
hereby appoint: my (relationship) ______________________________________
(appointee name) ____ as my attorney-in-fact (my "agent") to act for me and in my name in any way I could act in person to make any and all decisions for me concerning the care, medical treatment, hospitalization, and to require, withhold or withdraw any type of medical procedure for my animal(s), even though death may ensue. My agent shall also have full power to make a disposition of any part or all of my animal's body for medical purposes, authorize autopsy and direct the disposition of my animal's remains.

This Power of Attorney shall become effective on (start date) _____/_____/_____ and continue until: (check one) ❑(end date) _____/_____/_____, or ❑until further notice.

If any agent named by me shall die, become legally disabled, incapacitated or incompetent, or resign, refuse to act, or be unavailable, I name the following:

____

____

I am fully informed as to all contents of this form and understand the full import of this grant of powers to my agent.

____

____

Principal(s)

____

Print Name(s)

LIMITED POWER OF ATTORNEY FOR ANIMAL HEALTH CARE, Pg.2
FOR: ______________________________

ADDRESS: ____

The principal(s) has had an opportunity to read the above form and has signed the above in our presence. We, the undersigned, each being over eighteen years of age, hereby witness the principal's signature at the request and in the presence of the principal, and in the presence of each other; the day and year above set out.

____

Witness

____

Print Name & address

____

Witness

____

Print Name & Address

The foregoing instrument was acknowledged before me this _____ day of _____________ 20_____, by

____

____

Notary Public/Justice of the Peace
My commission expires: _____/_____/_____

This form developed in cooperation with Drs. Tomas and Rebecca Gimenez.

INDEX

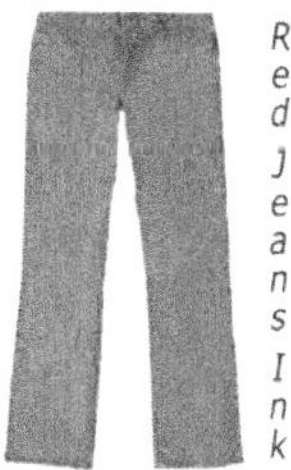

QUICK ORDER FORM

Website orders: www.redjeansink.com.
Postal orders: Send this form and your check to:

Red Jeans Ink, PO Box 881, Buffalo, NY 14201 ***or***
25 Phipps Street, Fort Erie, ON Canada L2A 2V3

Name: ____

Shipping Address: ______________________________________

City, State/Province, Zip: _________________________________

Country: ________________ **Telephone:** ______________________

Email address ____

I'd like to order the following:

_____ *Save Your Horse! A Horse Owner Guide to Large Animal Rescue* **Book**

_____ *Introduction to LAR for Horse Owners* "Teach It Yourself" **Class**

_____ *Horse Awareness and Safety* "Teach It Yourself" **Class**

_____ *Emergency First Aid for Pets* "Teach It Yourself" **Class**

_____ *Save Your Horse!* Book/LAR for Horse Owners Class **Combo**

_____ *Save Your Horse!* Book/Horse Awareness and Safety Class **Combo**

All "Teach It Yourself" classes contain a Powerpoint Presentation on a CD. Directions for setting up the class, plus additional information, are included.

Price per **Book**:	$29.95 US
Price per **CD**:	$49.50 US
Price per **Combo**:	$63.28 US

Shipping for single book/CDRom or Combo by Priority Mail:

US: $9.00 US
All other countries: $14.00 US

For multiple order shipping costs and bulk discounts please email us at info@redjeansink.com.

Made in the USA
Las Vegas, NV
05 April 2025